THE
OCEANS

THE OCEANS

A Deep History

EELCO J. ROHLING

PRINCETON UNIVERSITY PRESS
Princeton and Oxford

Published by Princeton University Press,
41 William Street, Princeton, New Jersey 08540

In the United Kingdom: Princeton University Press,
6 Oxford Street, Woodstock, Oxfordshire OX20 1TR

press.princeton.edu

ISBN 978-0-691-16891-3

Library of Congress Control Number: 2016960241

British Library Cataloging-in-Publication Data is available

This book has been composed in Gotham XNarrow & Sabon Next LT Pro

Printed on acid-free paper. ∞

Printed in the United States of America

1 3 5 7 9 10 8 6 4 2

For Mark and Sander,

and all others who were denied a future.

May all recipients of that greatest gift

safeguard it for future generations.

CONTENTS

**THE
OCEANS**

CHAPTER 1

INTRODUCTION

You may have been told that we know more about the Moon than about the oceans. It is a claim that I often hear from public relations staff, teachers, schoolchildren, and many others in casual chat at parties or receptions. But I work on ocean science for a living, and I am not convinced that it's true. Instead, it strikes me as evidence that most people have no real idea of how much we actually do know about the oceans.

I have been researching and teaching about the oceans for 30 years in the field of paleoceanography. In plain words, paleoceanography studies the ancient oceans and their changes through geologic time, before historical records. In practice it is interpreted to include the associated climate changes because there is a close interaction between the oceans and climate—you cannot research one and not the other. Well over 1000 researchers are currently active in this field worldwide. The number is several times larger if we include the wider ocean sciences, and especially if we include naval research as well. Although this still may not make it a large community, it has been remarkably productive in deciphering the natural, underlying rhythms and processes of ocean and climate change, against which humanity's impacts can be assessed.

A sound general awareness of this knowledge is essential to a well-rounded understanding of the Earth. As such, it is necessary to any meaningful discussion about matters of conservation and policy definition, and—indeed—to underpin election of knowledgeable

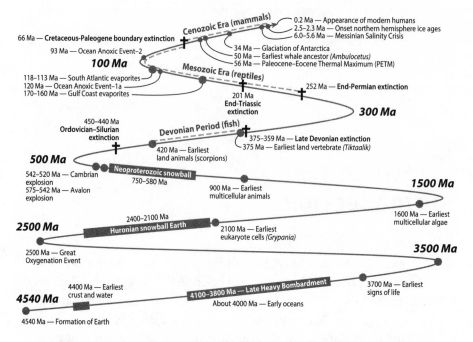

Figure 1. Timeline of events as discussed in this book. For a more formal representation of the geologic time scale, see figure 2. Ma stands for millions of years ago.

government representatives. The apparent lack of such awareness can only mean that my colleagues and I have fallen short; that we have insufficiently communicated our findings beyond the narrow circle of specialists. With this book, I hope to change that.

I venture here outside my scientific comfort zone, aiming to share our understanding of how the oceans have come to be the way they are today. Thus, I aim to offer a foundation upon which you, the reader, can build your own informed opinions about the challenges resulting from humanity's relentless drive to exploit the world ocean and its resources. To achieve this, we will travel through time from the first development of oceans more than four billion (or 4000 million) years ago, to the present. Along the way, we will stop at some of the most remarkable events and developments of life in ocean and Earth history.

We will see that life on Earth, including humanity, is closely intertwined with changes in the oceans and climate, no matter whether these result from natural variability or human impacts. These intricate

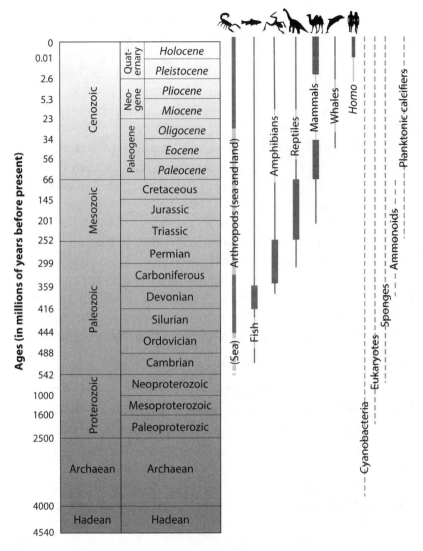

Figure 2. Geologic time scale with time ranges of major organisms discussed in this book. *Left column*, ages in millions of years ago. Note that the scale is nonlinear, for presentation purposes. From 4540 to 542 Ma, the *left-hand gray column* gives geologic eons and the *right-hand gray column* geologic eras. More detail is needed for the Phanerozoic eon, which spans from 542 to 0 Ma, so for that time the *left-hand gray column* gives geologic eras, and the *right-hand gray column* geologic periods (*regular font*) and relevant epochs (*italics*). *Homo* is written in italics as it refers to a specific genus. Bars indicate time ranges, with a continuous thin or dashed line for total range of named groups, and with thicker lines for periods of relative dominance. The arthropods—which include scorpions, spiders, millipedes, centipedes, insects, trilobites, crabs, lobsters, crayfish, krill, shrimp, woodlice, ticks, mites, etc.—have always been immensely numerous. Today, roughly three-quarters of all known animal species are arthropods.

connections between the oceans, climate, and life have been shaped and refined over four billion years. But we will also see that abrupt and dramatic adjustments which have taken place within this complex network, and may take place again, and that highly developed organisms seem to suffer the consequences most intensely. The explosive growth of humanity and our ever-increasing resource dependence make us especially vulnerable.

To gauge the potential consequences, we need to understand the natural variability that underlies human impacts. To learn about natural variability, all we can do is study the past before humans became important. To appreciate the scale and rapidity of human impacts, we need to compare and contrast them with a sound understanding of the underlying natural processes. Without the background knowledge, all else is conjecture. With the background, we stand prepared to make knowledgeable decisions for sustainable actions that will both give us what we need and preserve the system for posterity.

It won't surprise anyone who regularly watches the news that humanity's impacts on the oceans are multifaceted and profound. We pollute the oceans with medium- to long-lasting materials such as plastics, netting, radioactive waste, and dumps or wreckage laden with petroleum or chemicals. We disturb their food webs by overfishing, and by harmful fishing practices such as bottom-disturbing trawling and fishing with explosives. We overfeed (eutrophicate) them by pumping unnaturally large amounts of nutrients into their ecosystems by way of agricultural runoff, effluents, and detergents. We acidify them with our massive fossil-fuel-based carbon injection. And we cause unprecedentedly fast warming with our greenhouse-gas emissions, mainly in the form of carbon dioxide (CO_2) and methane (CH_4).

We have allowed all this to happen because of a deep-seated, historical notion that the world ocean and its resources are limitless, despite frequent warnings from modern scientific assessments that this is a serious mistake. Probably the worst part of it is the rapid acceleration of the problems. Some 500 years ago, the oceans still were pretty much pristine. By and large, they suffered only moderate change until the early 1800s, by which time the world's human population had reached one billion. Then things changed: large-scale whaling developed from about 1820, and global fish capture started to rise (today, it is well over 10 times higher than in the early 1800s). In the last two centuries, overfishing, pollution, eutrophication, acidification, and warming have risen sharply, on the back of fast human population rise, industrialization, globalization, and consumerism.

Some issues are more visible, like pollution, which gives rise to stronger public support for measures to address them. Other issues are all but invisible, like acidification, and their existence is hardly acknowledged outside scientific circles. But each issue has an element of finality if ignored—all require urgent attention. And because the seas know no borders, this requires globally concerted efforts. Even if a few nations won't join in, action by a majority will still alleviate much of the problem. Waiting for unanimity is nothing but a waste of time doing nothing.

Although they are serious issues, pollution and food-web disturbances fall outside the scope of this book which is concerned with the oceans through Earth history. Focus lies firmly on the other three key issues. We will encounter several events in ocean history that emphasize the urgency of curtailing eutrophication, to halt the spread and intensification of anoxic dead zones—zones in which decomposition of organic remains consumes all oxygen (anoxic means without oxygen). Past oceanic anoxic events reveal a devastating impact on ecosystems, while recoveries typically take thousands of years. The book also deals extensively with the nearly invisible "silent killer" issues of acidification and rapid warming. We will see along our journey through Earth history that such events had particularly devastating consequences.

Toward the end, I will draw on the book's geologic perspective to illustrate that the time for political debate and maneuvering is well and truly over—that a sustainable future requires immediate action. All signs indicate that without such action we are likely to rush headlong over a cliff into one of Earth's largest extinction events. Here, it is critical to emphasize that all documented past events measurably played out in the real world—not just in our minds, in a computer model, or on another planet. It is therefore impossible to dismiss or politicize them as fictitious, impossible, irrelevant, alarmist, or scaremongering. The way I like to look at it is that Mother Nature has left us this beautiful record, which shows that—when given time—she knows full well how to process our legacy, but that we are not likely to enjoy the way she resolves these problems. We can either heed the warning, or not. But we cannot play ignorant and pretend that no clear warning was issued.

*****

Human interest in the seas and oceans goes back to our very roots. Archaeological studies of the earliest modern humans, almost 200,000 years ago in South Africa, have revealed that we as a species have long been attracted to the sea and its seemingly limitless supply of food and shells. Along many of the world's coastlines, archaeologists have found large heaps of discarded shells that often show signs of cooking or breaking for consumption of the soft parts. Shells are also among the earliest items used in decorative art. So it's confirmed: we have always liked to collect shells on the beach.

There is good evidence that humans mastered the skills of rafting or boating at least as early as 50,000 years ago. The peopling of Australia involved the crossing of a few deep-ocean passages through the Indonesian archipelago, which never fell dry during even the greatest ice-age drops of sea level. Arguably, though, the rafting or boating skills may be even older. There are strong indications for an out-of-Africa migration of modern humans 60,000–70,000 years ago, across the southern Red Sea connection with the Indian Ocean, which again did not fall dry.

The enduring success of early humans in navigating the sea and exploiting its resources indicates that these people were carefully observing and studying the movements of the sea and the rhythms of life within it. We can safely presume that such enduring success could not be had just by chance. As any mariner will tell you, the sea is far too hostile and unforgiving for that. Over time, many civilizations and nations became heavily invested in seafaring, and trade, exploration, and research developed hand in hand. Ocean studies eventually became organized into oceanography, which spans the study of physics, chemistry, and biology of the ocean, and marine geology and geophysics, which focus on the geology and physics of the surface and subsurface of the seafloor.

Research in oceanography, marine geology, and geophysics has especially accelerated during the last century, triggered by wartime and cold-wartime needs for better ocean surveying and seafloor mapping. Consider the following examples from a very long list. First, much attention was given to understanding the impact of temperature and salinity (salt-content) contrasts in the oceans on the way sound waves travel, and on documenting these patterns in the oceans, with obvious implications for playing hide-and-seek with submarines. Arrays of underwater microphones, or hydrophones, were deployed to monitor enemy navigation, and after the end of the cold war many of these were retasked for research purposes. Second, there was a general push toward creating detailed maps of the seafloor, to support all manner of military and civilian purposes. Third, submarines beneath the Arctic sea ice routinely monitored ice-thickness variations above them to ensure that they could safely punch through the ice for communications or other operations. The ice-thickness records have since become available and now are invaluable in studies of Arctic sea-ice reduction over the past five or six decades. These initially military and increasingly science-driven research efforts have continued ever since, broadening out to include all aspects of relevance to the oceans and the ocean floor. Today, both fields of oceanography and marine geology/geophysics are thriving on a global scale.

Paleoceanography arose from marine geology, when researchers began to unravel past ocean changes by studying sediments sampled from the seabed. The HMS *Challenger* expeditions between 1872 and 1876 are commonly credited with the first systematic study of deep-ocean sediments. They are arguably paleoceanography's touchstone. At the time, however, only simple scoops of surface sediment were taken and studied. Systematic recovery of continuous sediment cores that penetrated one to two meters into the seabed started with the German South Polar Expedition of 1901 to 1903. Longer cores were obtained and studied by the US Geological Survey in the late 1930s. Routine recovery of relatively undisturbed sediment cores up to 10 meters' length became possible with the invention and first application of the Kullenberg piston corer, during the Swedish Deep Sea Expedition of 1947 to 1949 aboard the research vessel *Albatross*.

These developments in coring paved the way for the real emergence of paleoceanography as a separate research discipline in the early 1950s. The fathers of the discipline were the Swede Gustaf Arrhenius and Italian Cesare Emiliani (who did almost all his work in the United States). They were the first to apply sediment geochemical and microfossil stable oxygen isotope data in sediment cores to investigate circulation and temperature changes through time. An intense international research effort ensued that consolidated the discipline, with particular leadership in biogeochemistry by the American Wallace Broecker, and in the field of physical ocean-climate interactions by Sir Nicholas Shackleton in England.

As the discipline grew rapidly, it began to drive the development of new coring techniques. Core lengths of up to 70 meters became possible from the 1990s with the Calypso corer, designed and perfected by the French marine engineer Yvon Balut. Even longer cores were obtained using specialist drilling vessels. Scientific deep-sea drilling has been conducted since 1968 by a large international collaborative consortium that started out as the Deep-Sea Drilling Project (DSDP), then became the Ocean Drilling Program (ODP), and currently goes by the name of International Ocean Discovery Program (IODP). IODP remains the only research organization that is able to routinely undertake deep-sea drilling for science.

Despite these advances in scientific exploration, humanity's ever-increasing interest in the oceans was never just for research purposes. People's greatest interest is in exploiting their vast array of biological, mineral, and energy resources. The evolution of this interest has kept close pace with humanity's own development; arguably, the two are so tightly intertwined that one cannot be conceived of without the other. This relationship essentially stems from the vast—seemingly infinite—scale of the world ocean.

Ours truly is a blue planet, with more than 70% of its surface area covered by oceans. The world ocean is also very deep, reaching an average depth of about 3700 meters, and a greatest depth of about 11,000 meters in the Challenger Deep of the Mariana Trench between Japan and Papua New Guinea. The message is clear: the world ocean is enormous. This has given humanity the mistaken impression that the oceans are limitless, which underlies a range of our behaviors, from large-scale overfishing without heeding careful assessments of what can be sustained, to enthusiastic and unceremonious dumping of vast quantities of unwanted by-products of civilization. These by-products include short- and long-lived refuse, radioactive waste, outwash of chemical pollutants and nutrients, wreckages, heat (cooling water), concentrated salt from desalination plants, and so on.

Through it all, the world ocean has remained our loyal friend. Quietly and unassumingly, it has done us what may turn out to be the greatest favor of all: into its vastness, it has absorbed more than a third of humanity's total carbon dioxide emissions since the start of the industrial revolution. This has limited the rise of atmospheric carbon dioxide levels to the current value of 400 parts per million (ppm), rather than almost 500 ppm if the oceans had not done so. But we can't see it, for the outward appearance of the oceans has remained the same. It is only through specialist measurements that we have learned about the consequences of the carbon dioxide absorption. It has acidified the ocean waters, and ocean acidification has important implications for marine life. There are many examples of this in ocean history—we will discuss several in this book.

The world ocean is by far the largest reservoir of carbon within the actively exchanging system of atmosphere, water, and life, including dead matter. It contains 15 times more carbon than the atmosphere and all living and dead (soil) organic matter on the planet combined. In consequence, even small changes in oceanic carbon cycling will have large impacts on the atmospheric concentrations.

The oceans also sustain a rich and intricate food web, and humans have long relied on fish and other food supplies from the oceans. Increasingly, other ocean resources are being exploited as well, such as oil and gas, minerals, and metal ores from the seafloor, along with tidal, wind, wave, and heat energy. Even the water itself is becoming an important resource: desalination supplies fresh water for consumption and agriculture. And let's not forget the economic (if not environmental) benefits of the oceans' role as a receptacle for our waste, and as transport medium in global shipping.

In terms of energy, the world ocean dwarfs the atmosphere. The atmosphere may be some 60,000 meters thick (with three-quarters of its mass concentrated in the lower 11,000 meters), but the entire amount of heat it can hold only equals that of a 3.5-meter layer of ocean water. In other words, the so-called heat capacity of the world ocean is over 1000 times greater than that of the atmosphere. Exchanges of just fractions of that vast store of energy dominate the world's most dramatic weather patterns and cycles, including tropical and mid-latitude storms, monsoons, and El Niño.

The world ocean's vast heat capacity has moderated the rise of atmospheric temperatures over the past two centuries, by taking up more than 90% of the heat associated with global warming. Yet the addition of all this heat into the oceans has remained almost invisible because the oceans' great capacity to absorb heat means that only a minor rise took place in ocean temperature. That change is so small that it has been hard to measure accurately. Researchers have only managed to determine the enormous amount of heat energy stored in the ocean within the past decade or so.

While the oceans can effectively absorb heat, this process is not without consequences. Ocean heat uptake is related to ocean circulation, and consequently varies through time, which in turn affects the atmosphere. When the oceans absorb more heat, atmospheric warming slows down. When the oceans absorb less heat, atmospheric warming accelerates. Incidentally, this interaction can explain most—if not all—of the so-called "global-warming pause" of the past decade, which was more of a weak slowdown than a real pause. The oceans absorbed more heat than normal during that period, and this caused a slower rise of atmospheric temperature than might have been expected. In 2015, the ocean brought substantial amounts of that excess absorbed heat back in touch with the atmosphere, especially in the Pacific through the combined action of a strong El Niño and a remarkably warm surface-water blob in the North Pacific. Atmospheric temperature jumped up in response, making 2015 easily the warmest year since records began. The overall warming trend is highly consistent with the ongoing rise in CO_2 levels, with superimposed multiyear "wiggles" related to volcanic activity, and further superimposed short-term variability related—among other effects—to El Niño activity. Strong El Niño years commonly rank among the warmest years on record, but 2015 forms an exceptionally strong anomaly even by that standard. As the El Niño declined by mid-2016, the temperature anomaly diminished somewhat, but the underlying long-term trend of ocean warming remains indisputable regardless of such superimposed shorter-term variability.

With their uptake of the lion's share of heat associated with global warming, the oceans again helped us out in a very big way, in addition to the uptake of more than a third of our carbon dioxide emissions. An important question arises: How long can the oceans keep helping in this way? The answer is not known yet. But, eventually, the oceans' net uptake of carbon dioxide and heat will slow down, and a portion will be returned to the atmosphere in a persistent, long-term manner. Release of the heat energy will drive an increase in extreme weather, and release of the carbon dioxide will cause a strong acceleration of warming. One thing is certain: neither will be very pleasant. Studies of past changes in the oceans and their interaction with climate, as discussed in this book, provide valuable background for understanding the relevant processes and how they may affect the present and the future.

Although research activity strongly intensified during and after World War II, much remains to be explored and learned about the oceans. There are similarly large logistical and financial challenges in driving research that covers the oceans' great expanse and depths. It is striking that President J. F. Kennedy not only set up the space race by stating the ambition to land a man on the Moon, but separately also triggered an important increase in ocean research. He did this in his 1961 address to the US Congress, which included the statement that "Knowledge of the oceans is more than a matter of curiosity. Our very survival may hinge upon it."[1] Unfortunately for ocean research, support did not increase to the same level as that for the space race; as a result, much about the oceans remains unknown.

To illustrate the financial constraints on ocean research, consider that the use of a major research vessel, typically between about 70 and 120 meters in length, costs upward of 20,000 US dollars per day, plus fuel and salaries. And you may magnify this by many times for a research icebreaker, or—even worse—for a deep-sea drilling vessel.

---

1 John F. Kennedy, "Letter to the President of the Senate on Increasing the National Effort in Oceanography," March 29, 1961, Gerhard Peters and John T. Woolley, *The American Presidency Project*, accessed February 9, 2017, http://www.presidency.ucsb.edu/ws/?pid=8034.

If the ship is going to work in the heart of a major ocean basin like the Pacific, then it has to spend more than two weeks just going out to the study region and the same again for returning to port, plus at least three weeks for research on location. That adds up to an eight-week expedition, for just three weeks of actual work. Any bad weather on location or equipment breakdown, and the proportion of time for work gets reduced even further. In all, we're talking about well over one million US dollars per trip, and all it gives us is a snapshot in time of the state of the ocean in a region of very limited size. On top of this, several years of research are needed to work up the information and materials collected, adding further costs, especially in person-hours. Some four or five of these trips are possible per ship, per year, and the costs quickly explode beyond most governments' budgets for ocean studies. In the absence of a major focused resolve like that involved with the race to land on the Moon, this strongly limits the rate at which we can gather scientific information about the oceans and seafloor, especially in remote areas.

Another reason why much remains to be learned about the oceans is that they present a very challenging environment to study, with unpredictably dangerous conditions, salt that attacks all equipment, enormous pressures at depth that require specialist engineering, and so on. It takes continuous specialist equipment design and engineering. And it takes guts. I have endless respect for colleagues who volunteer to dive thousands of meters below sea level in submersibles, called bathyscaphes. These may be specially designed craft, but it's still a very cramped little bubble of space that people take to go into an enormous pitch-dark void, down to as much as 10 times greater depths than even the deepest-diving military subs can handle. There, pressures are up to 1100 times atmospheric, more than enough to instantly crush the craft and everything inside it into a pulp at the first hint of a problem. It's not so different from the technological and personal challenges of going into space. In 1960, the first manned dive reached the greatest depth of the world ocean in Challenger Deep (almost 11,000 meters), by Jacques Piccard and Don Walsh aboard the bathyscaphe *Trieste*. This was only a year before the first manned spaceflight by Yuri Gagarin. Since then, just one further manned dive has reached that depth, by James Cameron aboard *Deepsea Challenger* in 2012. Other very deep dives have been made with the Chinese *Jiao-long*, which reached about 7000 meters in 2012, and the Japanese *Shin-kai 6500*, which can go to 6500 meters' depth and made the world's first live broadcast from 5000 meters in 2013.

In recent decades, satellite observations have revolutionized the mapping of surface properties (such as temperature, salinity, currents, plankton blooms, and ice cover), and of the sea-floor topography, or—as it is technically known—ocean bathymetry. But satellites cannot look at water properties or biology below the surface; for this, there is little choice but to go out on ships. In addition, special equipment is commonly needed, which rules out the option to do research from "ships of opportunity," which are merchant ships that travel the oceans anyway. This is especially true for research that relies on the lowering of measurement equipment into the deep sea using super-strength lightweight cables because of their enormous length, on conducting cables that are needed to electrically activate equipment or collect data, on specialist deployment of manned or unmanned submersibles, and on the deployment and recovery of sediment cores that are up to 70 meters long or of drill-core sequences that can reach many hundreds of meters in length. For such activities, dedicated research vessels remain the dominant option.

Very recent developments are beginning to challenge the research vessel's position of dominance. These include an array of unmanned measurement devices, such as autonomous underwater vehicles, underwater gliders and drifters, floats, animal-borne sensors, expendable sensors that are launched from ships or dropped from planes, and seafloor-mounted and seafloor-lander measurement systems. Integration and coordination of much of that technology occur through a bewildering array of major (inter-) national programs that include the Ocean Observatories Initiative, the Open Ocean Observatories, the Global Ocean Observing System, the Dense Oceanfloor Network System for Earthquakes and Tsunamis, and a variety of buoy- or cable-based infrastructure programs with exotic acronyms such as ORION, NEPTUNE-Canada, ARENA, and ESONET, which would fill the page if written out. Suffice to say that there is a lot of activity in this development of the future of marine science. If you like robotics, sensor technology including nanotechnology, "big-data" approaches, "thinking outside the box," and—in general—a challenge, then here's a great opportunity to make a difference.

*****

Past studies into the oceans and their history have brought a wealth of knowledge that serves as a sound foundation for continued ocean research. In this book, I will share key elements of this knowledge, by going through a history of major changes in the oceans (see figure 1). Along the way, we will explore the drivers of ocean properties and circulation, of life in the ocean and so-called biogeochemical cycles, and of physical and chemical changes in the oceans. Some technical terminology is unavoidable. I will do my best to minimize its use, and where it cannot be avoided, I will explain the terms in the text.

Progressing through the discussions, I introduce a lot of information about geology and geophysics, evolution, and mass extinctions. This is because the oceans cannot be seen separately from the wider developments of the planet and its changing continental configuration, and because evolution and extinction of life have been strongly tied with changes in the oceans. In addition, these broader components of the Earth system are important to the forward look that is presented in the final chapter.

I hope that this book will demonstrate that the world ocean is vast but finite, and that everything within it is interconnected in such a way that proper planning and governance are needed to prevent us from destroying the system. It is easy to gaze out over the seas and think that such a thing could never happen, that humanity is too small to affect this massive system. But one couldn't be more wrong. What's more, if we overstep the limits, the consequences will be irreversible on human time scales. Consider, for example, that sea-level rise will be with us for centuries, that a collapsed ecosystem will take decades or longer to make some sort of recovery and the end result will be forever different from the original, that ocean acidification—especially once it has penetrated into the deep sea—will need thousands of years to be brought back to normal, and that heat stored in the surface ocean will continue to feed powerful weather extremes for decades.

Enough already, we're rushing ahead of ourselves—let's start at the beginning.

CHAPTER 2

ORIGINS

To begin the discussion of how the oceans formed, I must take you back to the birth of Earth itself, along with the rest of our solar system. Earth formed just under 4.6 billion years ago. It is important to let that number sink in. If we scaled Earth's 4.6-billion-year history to just 24 hours, then the entire 1 billion years' period that multicellular life has existed on Earth would span roughly 5 hours and 20 minutes, the 66-million-year "age of mammals" since the demise of the dinosaurs only 20 minutes, and the 200,000-year duration of modern human existence less than 4 seconds (see figures 1 and 2). Our settled, non-nomadic lifestyle started about 11,000 years ago; this scales to an insignificant one-fifth of a second on our clock, which literally is only half a blink of an eye.

Oceans formed early on planet Earth; we have good evidence of their existence by about four billion years ago, although their shapes have been continuously shifting owing to movement of the continents. It is good to reflect on this for a bit. Since starting at university in 1981, I have seen the age estimate for the earliest oceans creep further and further back in time, as new data became available. It never ceases to amaze me how large bodies of water had established themselves so very early in Earth's history, especially given the violent processes that shaped early Earth. But the evidence is there, so let's have a look at this history. This chapter starts with an account of the planet's formation, followed by the appearance of the oceans and their shape-shifting nature. Then we discuss where the water came from, why it is salty, and how ocean circulation is driven. Finally, we consider the impacts of life on the cycles of oxygen and carbon in the oceans and the wider Earth system.

BUILDING A PLANET, SHAPING THE OCEANS

Both theory and observations of planetary formation around other stars tell us that Earth formed through a process of gravity-driven clumping of matter in a disk of gases and molecular dust that was spinning around the forming Sun. At first, microscopic particles and little specks of dust clumped together, and collisions with others made these grow into larger lumps to form a "pebble ring." Eventually, the lumps clumped together into the larger masses of so-called "planetesimals" and even larger protoplanets, which are smaller planet-like bodies. Many such bodies formed in the disk of dust that was spinning around the Sun. The increasing gravitational attraction of the ever-growing protoplanets pulled in more and more matter, which made them grow even faster. The planets we see today are the survivors that have vacuumed up virtually all other matter in their orbital paths. Earth is the survivor in its own specific orbital band around the Sun.

All elements that we know today on Earth were gathered from space- and stardust in this manner, except for the more than 20 synthetic elements that humans have produced during the last century. In orbits close to the Sun, temperatures are much higher than further out in the solar system. As a result, lighter elements were vaporized in the inner solar system, and these volatile vapors were—for a large part—blown into the outer solar system by the solar wind, the Sun's powerful emission of charged particles. Far away from the Sun, where the solar wind is weaker and temperatures are lower, the gravity of the outer planets captured the vapors of light elements, which at the low temperatures condensed into denser molecules. Thus, heavier elements came to dominate the rocky planets of the inner solar system (Mercury, Venus, Earth, and Mars), while lighter elements collected on the outer gas-giant planets.

On Earth, the earliest 50 to 100 million years were particularly violent. Earth, essentially, was a seething cauldron of molten rock without a permanent crust. There were frequent impacts of smaller and larger bodies. Within that earliest period, a remarkable event happened that led to formation of the Moon. Radiometric dating and comparison between the chemical compositions of Moon, meteorite, and Earth rocks indicate that the Moon formed with a bang. Around 4.5 billion years ago, Earth was dealt a glancing blow by another newly forming planet, Theia, which was roughly the size of Mars. The two planetary bodies melted together, but the collision also launched a major cloud of debris into orbit around the Earth. This debris eventually clumped together and—under its own gravity—smoothed out into the spherical shape of the Moon. Incidentally, the impact is also thought to be responsible for knocking Earth's axis into a position that is today tilted by 23.5 degrees. Without this, there would have been no seasons.

Another, less visible but equally important, event in Earth's seething earliest phase is known as the differentiation event, or "iron catastrophe," which completely changed the initially homogeneous composition of Earth. This event happened around 4.5 billion years ago, when the planet had grown large enough for pressure to drive temperatures in the interior above 1000°C, the point at which rocks melt. Then, denser (metal-rich) materials sank to the center of the planet, and less dense (rocky) materials rose toward the surface. The sinking dense materials formed Earth's nickel-iron alloy core, the planet's inner 3500 kilometers or so. The lighter materials that rose up formed the less-dense rocky mantle, the planet's outer 2900 kilometers.

The formation of Earth's core transformed conditions on Earth's surface. This is because it created the right conditions for development of the planet's magnetic field, which originates from movements in the outer layers of Earth's core. It had been known for a while that the magnetic field was already operational by about 3.5 billion years ago, and very recent work has brought that back to before 4 billion years ago. The magnetic field is Earth's only real protection against the solar wind, which was stripping gases from the earliest atmosphere before the magnetic field had started up. Thus, the differentiation event is thought to have been critical for reducing the loss of light elements from the atmosphere. Without it, Earth might have ended up without hydrogen, and thus without water. And over time, many heavier gases would also have been stripped off by the solar wind. Mars is thought to have started out with a magnetic field, but—being much smaller than Earth—to have cooled enough for its magnetic field to die at around four billion years ago. It subsequently lost almost all of its atmosphere and surface water. While this often-used explanation for retaining an atmosphere by presence of a magnetic field sounds plausible, some further thought suggests that things may be a little more complicated. Venus has no magnetic field and is closer to the Sun, yet has a very well-developed atmosphere. Venus and Earth have similar sizes and masses, while Mars is much smaller—hence, gravity may have been equally or more important for retaining an atmosphere than a magnetic field, especially when the gases concerned are heavier gases, like the dominant CO_2 on Venus.

While vacuuming out its orbital path around the Sun, early Earth was battered by comets, meteorites, asteroids, and even protoplanets like Theia. In addition, it was hit by many bodies that had been kicked out of orbit by the other planets, into trajectories that eventually collided with Earth. An especially intense period of asteroid and comet impacts, the Late Heavy Bombardment, occurred between 4.1 and 3.8 billion years ago, as evidenced by Moon rocks and asteroid fragments, as well as by numerous craters on the Moon, Earth, Mars, and Mercury. Despite this onslaught, Earth's surface still managed to cool down quickly by heat loss to space.

At the planet's position in the earliest solar system, temperatures of 250°C to 350°C would be expected, but the energy from the intense early clustering and impacting of gases, dust, pebbles, comets, meteorites, asteroids, and planetesimals had pushed temperatures up well above the melting temperature of rocks. Still, work on radiometrically dated crystals of the mineral zircon from western Australia's Jack Hills has demonstrated that, as early as about 4.4 billion years ago, Earth's surface had not only cooled sufficiently to form early crust (likely below 1000°C), but even enough to allow for the presence of liquid water, which at modern atmospheric pressure would mean that temperatures had dropped below 100°C. However, at higher pressures this value is higher, and we don't really know how dense the early atmosphere was. So 100°C is a lower estimate; true temperatures may have been double that value if the early atmosphere was very dense.

Now that clouds and rain had appeared on the scene, oceans began to develop. Large portions of the planet were covered with water by about four billion years ago. This is somewhat unexpected because, at this time, Earth's fiery birth phase had rapidly settled down, and the Sun was only about 70% as strong as it is today. After their first bright ignition, stars like the Sun shine more weakly and then grow gradually more intense with age; we refer to this as the "faint young Sun." With only 70% of the modern energy coming in from the Sun, early Earth should have been covered by ice, not water. Somehow, Earth's early atmosphere must have retained heat more effectively than the modern atmosphere. Indeed, reconstructions of the early atmosphere's composition suggest high levels of carbon monoxide and dioxide, water vapor, and later also methane, whose greenhouse-gas characteristics would have caused efficient heat retention. There was no free oxygen yet, as we will see later.

So, cooling of early Earth allowed surface crust to form, about 4.4 billion years ago, and surface water appeared close to that time as well. With water around, weathering and erosion would have become important. Weathering is the chemical breaking up and dissolution of minerals. Erosion is the physical transport of fragments of material by a transporting agent, typically water, ice, or air. The transported fragments are known as sediment, and when sediments settle and are deposited, this forms sedimentary deposits. Compression and binding together of sedimentary deposits by natural cements lead to the formation of sedimentary rocks like sandstone, limestone, and mudstone. Sedimentary rocks started to appear close to 4.4 billion years ago, and some of these are still visible on fragments of early crust that have survived until the present.

By about four billion years ago, and likely even earlier, plates of early crust carrying sedimentary and volcanic rocks were moving around, similar to Earth's tectonic plates today. This movement is known as plate tectonics, and we will investigate it in more detail below. For reasons that go beyond the scope of this book, plate tectonics is needed to maintain the heat-flow motion in Earth's outer core that generates the magnetic field. Therefore, the recent discovery that the magnetic field existed before four billion years ago strongly suggests that plate tectonics was also operational before four billion years ago. It caused colliding plates to become buckled, and mountain ranges to be pushed up, like crumple zones on a car wreck. Uplift of air against mountains focuses rainfall, and this caused extra erosion and weathering. Combined, these processes led to the deposition of sedimentary units in valleys, and in lake and ocean basins.

Gradually, larger complexes formed from crustal plates that crashed into each other and stuck together, including their sedimentary and volcanic rock units. This gave rise to the formation of extremely ancient complexes of several tens of kilometers thick, geologically diverse crust, which we call "cratons." The cratons formed the earliest continents, and they can still be recognized at the hearts of all of today's major continents.

Oceans may have appeared at least four billion years ago, but they certainly did not look like they do today. The process of plate tectonics has caused continuous changes in the distribution of land and water throughout geologic time. This shape-shifting behavior is so fundamentally important to ocean history that we need to digress a little and consider what the oceans look like today, then what has caused their shapes to change through time, and finally how researchers managed to find out about this.

Today, the oceans are on average about 3700 meters deep, and their greatest depths are found in the trenches that are associated with subduction zones. Let's review the main features of the ocean's underwater topography, or bathymetry (figure 3).

Surrounding the continents are the continental shelves, where the edges of the continents continue under water. Typically, these shelves are less than 100 to 200 meters deep. Sometimes a continental edge has only a very small shelf area, and in other cases the shelf can be massive. For example, there are extensive shelves between Australia and New Guinea, between the islands and peninsulas of Southeast Asia, between Argentina and the Falkland Islands, between Alaska and Siberia, extending into the Arctic Ocean from the Eurasian continent, and connecting northwestern Europe with the United Kingdom and Ireland. These large shelf areas are shallow enough that they fall dry during ice ages, when such a large buildup of continental ice sheets occurs that global sea levels drop by 100 to 140 meters. The shelves listed above then form land bridges or major landmasses that connect places now

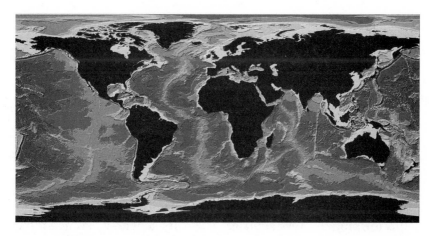

Figure 3. Global ocean bathymetry. Depth increases from light gray to dark gray, with categories shown in steps of 0–200 m, 200–1000 m, 1000–2000 m, and then every 1000 m until 10,000 m. A 3-D effect is achieved by image illumination from the northwest. Map created using free vector and raster map data from http://www.naturalearthdata.com.

separated by the sea. Some of these landmasses have even been named, such as Sahul (Australia-New Guinea), Sundaland (Southeast Asia), Falkland Plateau (Argentina-Falklands), Beringia (Alaska-Siberia), and Doggerland (northwestern Europe–United Kingdom–Ireland).

Incidentally, when I was a little boy in the Netherlands, bottom-trawling fishermen frequently used to "catch" mammoth skulls, tusks, and bones in the North Sea, which dated from ice-age land phases of Doggerland. Call me nerdy, but this used to fascinate me to no end, just like hunting for fossil seashells in the mountains (which I could only do during holiday trips because there are no mountains in my native country). These were important contributions to my motivation to get into geology and past environmental reconstructions.

Where the shallow shelf drops away to greater depths of a kilometer or more, we speak of the continental slope. Here, the transition occurs between thicker continental crust and 5 to 10 times thinner oceanic crust. These transitions may be simply where oceanic crust is stably attached to continental crust, as around most of the Atlantic Ocean; we call these passive plate margins because they are generally quiet for earthquakes and volcanoes. In other places, most notably but not exclusively around the Pacific, the transition is a subduction zone, where the shallow continental shelf suddenly drops away sharply into a trench. These are the active plate margins, with lots of earthquake and volcanic activity. Note that subduction does not have to be a process between continental and oceanic crust. There are many places with subduction of ocean crust beneath ocean crust—for example, at the Caribbean Island arc, and at the Mariana ridge and trench system.

Having dropped down the continental slopes, either smoothly or through a trench, and moving further out, we arrive on the vast and almost flat abyssal plains of the deep sea. These are the regions where ocean crust has moved away from the hot spreading ridges. Cooling and absence of widespread mantle upwelling then allows the crust to settle, forming the abyssal plains. These are generally between three and six kilometers deep, and cover about 50% of the Earth's surface area—a huge proportion relative to the total ocean's 72% of the Earth's surface area.

Especially in the Pacific, lots of active and ancient volcanoes poke up from the abyssal plains. As the crust moves away from the spreading center, it settles to greater depths, and volcanoes jutting up from the crust are often submerged as their growth slows down and they fail to keep up with the general sinking of the crust. Other volcanoes always remain under water—that really doesn't stop volcanic activity. These volcanoes create underwater mountains, or seamounts. There are about 100,000 seamounts that protrude more than 1000 meters above the abyssal plains that they stand on, of which about half are located in the Pacific Ocean.

Volcanoes in tropical latitudes that reach to, or rise above, sea level rapidly become colonized by corals. Reefs build up around such islands. These reefs grow upward as the volcano sinks or sea level rises, as long as things are not moving too fast. The Pacific is littered with active and ancient volcanoes. As the volcanoes die out, they no longer grow, but still keep gradually sinking with the crust. Reefs that had established themselves on these ancient volcanoes keep on growing, trying to keep up with the relative rise of sea level. After a (long) while, all that can be seen at the surface is a coral atoll. Some of such islands have submerged too fast, and the reef could not keep up. That then forms a "guyot," a flat-topped underwater mountain, or seamount. The Emperor seamounts, an ancient extension of the Hawaii volcanic chain, contain many guyots. But, equally, many seamounts and guyots stand in isolation, without being organized in chains.

Further along the abyssal plain, we eventually end up again at the midoceanic ridges where new oceanic crust is formed, as we will discuss next. The ridge axis is frequently offset in a sideways manner by transform faults that extend at more or less right angles to the ridge axis. Transform faults are often very long, and very active. Sometimes, they extend onto land, as in the case of the San Andreas Fault in California. Another major transform fault that can be observed on land is the Alpine Fault of New Zealand. Transform faults are usually associated with intense earthquake activity, which is of a shallow nature—within the crust—because there is no subduction. Earthquake activity of the San Andreas and Alpine faults is familiar to all from the news. Another transform fault that can be observed on land is the so-called Dead Sea transform, which is infamous as one of the deadliest faults on Earth. An earthquake along this fault in 1138 killed some 230,000 people. But transform faults are active all around the world, also when they are deep under the ocean. The 1755 Lisbon earthquake and tsunami, which killed an estimated 60,000 people in Lisbon alone, resulted from movement along a transform fault in the deep eastern North Atlantic.

Midoceanic ridges and subduction zones are critical to the shape-shifting nature of ocean basins. In an endless cycle, spreading at mid-ocean ridges opens up ocean basins, while removal of crust at subduction zones closes other ocean basins. These processes occur at roughly the rate at which your fingernails grow. That may be slow, but it is directly measurable using modern lasers and global positioning systems.

Mapped for the first time in the North Atlantic by the American geologists Marie Tharp and Bruce Heezen in 1957, and by the same team on a global scale in 1977, midoceanic ridges are now known to form the longest mountain chain on Earth. Although hidden almost entirely under water, the total length of spreading ridges amounts to some 40,000 nautical miles (1 nautical mile is 1.8 kilometers) (see figure 3). The ridges are not low, either; they reach up to 4.2 kilometers above the deep ocean floor. At such lengths, even slow spreading has big consequences when given enough time, and the great depth of geologic time provides plenty of that. The result is that entire ocean basins open and close over time scales of hundreds of millions of years, moving the continental plates around the planet. Sometimes continents crush an ocean basin away between them, and then crash into each other and merge, while pushing up vast mountain ranges; the collision between India and Asia squashed an ancient ocean into oblivion and pushed up the Himalayas. Sometimes continents tear apart, as we see happening today in the East African Rift. When such rifts connect with the oceans and fill up with water, they form young ocean basins like the actively rifting Red Sea.

At the midoceanic ridges, rifting and upwelling of magma cause formation of new oceanic crust that typically is only about five kilometers thick. This process is associated with lots of earthquake activity, and because the crust is very thin in these regions—after all, the underlying magma wells up all the way to the surface—the earthquakes are generally shallow earthquakes, with depths of generally less than 30 kilometers. We can observe what happens at spreading centers not only underwater, with submersibles, but also on land in Iceland, which is a part of the mid-Atlantic ridge that sticks out above sea level (Hawaii is not the same; it is a hot spot that the Pacific oceanic plate slowly moves over, creating a volcanic chain). It is important to note that, irrespective of what their name may suggest, midoceanic ridges are not always found in the middle of every ocean basin. In the Pacific Ocean, the ridge lies toward the eastern end of the basin, and in much of the North Pacific it even cozies up with the North American continent. In the Atlantic Ocean, and to some extent in the Indian Ocean and Southern Ocean, the ridges lie closer to the middle of the ocean basin (see figure 3).

Apart from new crust generation, midocean ridges contain another feature that must be highlighted. Because of the thin crust, hot mantle is not far removed from ocean water, and this allows geothermal heat from the underlying mantle to heat up water that seeps into cracks in the ocean crust. The heating is accompanied by strong chemical weathering of the typically basalt-like crustal rocks. Inside the cracks, the water thus becomes hot, pressurized, and loaded with dissolved mineral components, and the chemical reactions also consume all its oxygen. The resulting superheated and chemically loaded water escapes from the crust in the form of hot-water, or hydrothermal, seeps (a gentle process) or vents (like an underwater geyser). The latter show up as dramatic white or black smokers, depending on their temperature and mineral composition. Hydrothermal vents have not been known for very long; the first active system was discovered as recently as 1977. Regardless, they have rapidly gained importance for two reasons; the first is commercial interest, and the second has to do with unique biological adaptations around them.

Along with dissolved minerals, hydrothermal flows are loaded with metal complexes. Their deposits contain a wealth of precious metals. Fossil hydrothermal systems have long been metal-mining targets, and active hydrothermal systems in the oceans have recently become targets for deep-seabed exploitation. This sets up a challenge of reconciling commercial interests with conservation of a unique ecology that is found around vents. This ecology is fueled by oxidation reactions that take place as vent waters with their dissolved mineral content hit the overlying oxygenated deep-sea waters. The vent ecology then relies on specialist microbes that can form organic matter using the energy released by oxidation reactions of, especially (but not only), hydrogen sulfide. As such, these microbes practice chemosynthesis, instead of the photosynthesis that we are used to seeing where plants use energy from sunlight to generate organic matter, which will be detailed a bit later in this book. Using chemosynthesis allows the microbes at hydrothermal vents to thrive perfectly well in the pitch-dark deep sea. Specialized ecosystems—including worms, shrimp, crabs, and so on—have evolved around this microbial chemosynthesis at deep-sea vents, in a similar way to how the complex ecosystems we are more familiar with have evolved around photosynthesis by plants and algae. As mentioned, we will get back to this later, when discussing the evolution of life.

Creation of crust at spreading ridges means that there must be destruction of crust in other places, since Earth is not significantly gaining or losing mass or size. Destruction of crust happens in subduction zones, where one plate dives underneath another, slipping into the upper parts of Earth's mantle. In these regions, we see deep trenches in the ocean bathymetry. Subduction generates a component of pull on the subducting (diving) plate, which combines with the push from spreading ridges to drive the creep of continental plates around the world. Subduction zones are very active regions for earthquakes, which develop all along the subducting slabs down to depths of as much as 700 kilometers. Deeper than that, earthquake activity sharply decreases because temperatures get high enough for deformation of the subducting slab to occur in a more plastic, or ductile, manner (like very tough putty), rather than through brittle breaking. We commonly see intense volcanism near subduction zones as well. Key example areas are Japan, Indonesia, New Zealand, Chile, western Central America, the Cascadian Mountains of the northwestern United States, and the Aleutian Islands—in other words, the Pacific "ring of fire," where Pacific oceanic crust dives underneath adjacent plates. Subducting plates carry sediment and water with them into Earth's hot mantle. This intensifies the formation of low-density melt and degassing of, mainly, CO_2, and these materials then bubble up through the overlying crust and drive intense and often dramatically explosive volcanicity.

The endless cycle of plates moving along Earth's surface has become known technically since 1953 as plate tectonics, but was previously known by the beautifully descriptive term of continental drift. It was put forward in a first comprehensive theory by the German geophysicist Alfred Wegener in the period of 1912 to 1915, although suggestions of continental movements around the planet had been around since 1596, and more specific inferences since the mid-1850s. The name of the process was changed because it was recognized not only that the continents were moving, but that they formed part of larger tectonic plates that include giant swaths of oceanic crust as well. The process of plate tectonics not only causes the shapes and sizes of ocean basins to slowly but unstoppably change through time. It also creates and destroys landmasses, continental connections, mountain ranges, volcanic centers, and so on. Plate tectonics is driven by the flows of Earth's internal heat engine. It is tempting to think that this heat engine relies on primordial heat from Earth's formation, but that heat is long gone already—if that had been the only source of heat, then Earth would have frozen solid very long ago. Instead, Earth's internal heat engine is maintained by radioactive decay of elements within the planet's interior.

To understand how researchers figured out the nature of plate tectonics, we need to start at the so-called "constructive" margins of the tectonic plates, where new crustal material is formed: the midoceanic ridges. Although we can today measure spreading rates with high-tech equipment, as mentioned above, this is not how the movements were first discovered. Instead, we got our first details about the spreading process from measurements of the magnetic field as recorded in oceanic crust along transects directed outward from the ridge axis. These measurements were made aboard ships that were towing magnetometers above the ocean floor to record magnetic field variations.

Earth's magnetic field has reversed regularly on time scales of 100,000 to a (few) million years, throughout geologic time. When that happens, magnetic north becomes magnetic south, and vice versa. As the magma that upwells at midoceanic ridges solidifies to form new oceanic crust, magnetic minerals inside it align themselves with the magnetic field direction that exists at the time. A magnetometer that is towed over crust with minerals aligned to a normal magnetic field direction (same as today's) will record that the modern magnetic field intensity seems slightly amplified. When the magnetometer is towed over crust that formed with minerals aligned to a reversed magnetic field (opposite to the present), it will find that the magnetic field intensity seems slightly weakened. In this way, magnetometer surveys from the ridges outward have mapped out a banding of ocean-crust polarity, from normal at the ridge today, to reversed, normal again, reversed, normal again, and so on. This banding was first discovered in 1955. But, as measured, it was just a pattern of magnetic polarity bands of different widths—there was a need to determine ages to go with that pattern, so that spreading rates could be calculated.

Radiometric dating of magnetic polarity reversals has been obtained from samples of volcanic lava sequences that are exposed on land. This type of dating is based in using the known rate of radioactive decay of certain elements that are included in the rocks. Applied to the magnetic reversal series measured for the ocean floor, such dating offers accurate ages to go with our information on spreading distance; thus we can calculate the rate of spreading of the ocean basins. The youngest (modern) ages are found at the ridge itself, where spreading and magma upwelling are happening as we speak. And the ages get older as we go away from the midoceanic ridges. This age increase is symmetrical in both directions from the ridge. In this way, the oldest ocean crust in the South Atlantic has been dated at about 150 million years old, and in the North Atlantic at about 180 million years old. These oldest crust segments are attached to the continents at the western and eastern outer margins of the basins, and they reflect the age at which the ocean basins started to rift open. This reveals that the North Atlantic started to open some 30 million years earlier than the South Atlantic Ocean.

Similar assessments have been made for the other ocean basins, including age determinations that were assisted by dated changes in fossil associations contained within the sediments. Thus it was found that the Southern Ocean consists of several opening sectors, which suggest that Africa started to separate from Antarctica at around 150 million years ago, Australia from Antarctica at around 60 million years ago, and South America from Antarctica only 30 or 40 million years ago. The Indian Ocean is complex too. A rift began to develop between Africa and Madagascar roughly 150 million years ago. Madagascar at the time was part of a plate with India and Australia. India and Madagascar started to rift away from Australia at around 120 to 130 million years ago, followed by an onset of rifting between India and Madagascar roughly around 90 million years ago.

Based on a concept of slow and steady spreading, we might expect that the very wide Pacific Ocean contains some extremely old oceanic crust, but it doesn't. In the Pacific, the oldest crust that can be found seems to date back to at most 200 million years ago because the old margins of the Pacific have been removed. In the Pacific, the margins of the spreading plates are no longer intact, in contrast to most of the separation margins discussed above. All around the Pacific, there are subduction zones where Pacific plates have detached from the continents and are now diving underneath them into the mantle; the Pacific margins are destructive plate margins. There still are active spreading ridges in the Pacific, where new plate material is being formed. But all around the margins of this mighty ocean, Pacific plate is being gobbled up in the mantle, where it eventually melts. The net result is that the Pacific Ocean basin is slowly closing—its contraction is making space for the expansion of other ocean basins. Looking back through time, we see that some of the Pacific margins descend from those of an extremely ancient ocean; namely the

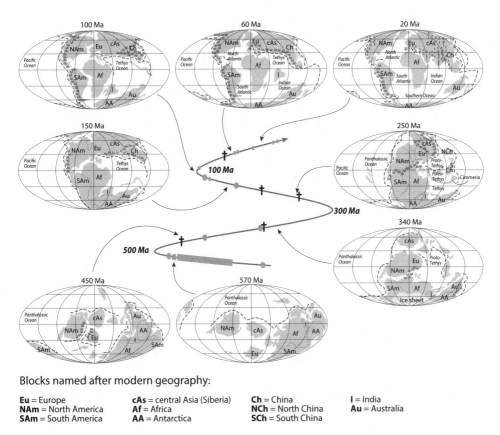

Blocks named after modern geography:

Eu = Europe	**cAs** = central Asia (Siberia)	**Ch** = China	**I** = India
NAm = North America	**Af** = Africa	**NCh** = North China	**Au** = Australia
SAm = South America	**AA** = Antarctica	**SCh** = South China	

Figure 4. Schematic of plate-tectonic developments during the last 570 million years, for which reasonable data exist. The maps are shown in relation to the timeline of figure 1.

Panthalassic Ocean that spanned the entire world around a single supercontinent called Rodinia, more than a billion years ago. Subduction at the eastern edge of the Pacific Ocean, along North and South America, can be traced back to that age or even earlier. Along the Asian edge, subduction started some time between 510 and 650 million years ago (figure 4).

Although we have seen that there is no oceanic crust left from those times, we still know quite a bit about these old dates and movements, thanks to ancient crustal components within the continents that date back to several billion years ago. As the continents were shifted around by plate tectonics, volcanic and other deposits that

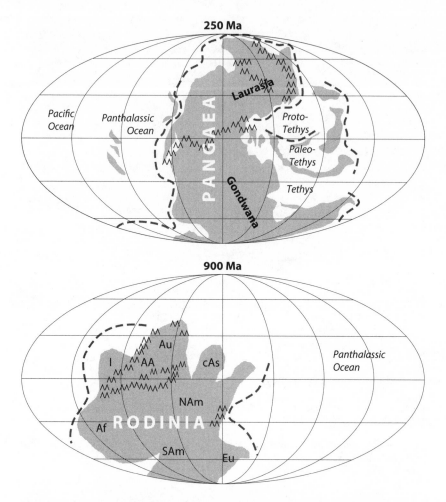

Figure 5. Schematic summary maps of the main supercontinents of the past one billion years. AA, Antarctica; Af, Africa; Au, Australia; cAs, central Asia (Siberia); Eu, Europe; I, India; NAm, North America; SAm, South America.

record magnetic alignments have formed on them at different times through geologic history. The combined deposits provide a record of changing continental positions relative to the Earth's magnetic field alignments. Tracking these yields a picture of plate-tectonic movements during very ancient times, even in the absence of any ocean crust (see figure 4). Summary outlines of past supercontinents are given in figure 5. In general, information is scarce for times before about 600 million years ago, notably because there are no fossils to help with age assignments for those times, and reconstructions keep changing as a result. I have not included detailed maps of these older times, as my sketches would be outdated very soon. If you are interested, then up-to-date plate-tectonic reconstructions may be found easily on the Internet from specialized sources (also for more ancient times; for example, using a search for "Rodinia").

WATER, SALT, AND CIRCULATION

All this tells us why ocean water is where it is, but it doesn't tell us how that water got there in the first place, or why it is so salty. Solving the riddle of the origin of Earth's water and atmosphere is important because understanding how Earth became what it is will tell us much about the chances that other planets in other solar systems may have developed in similar ways. These might just have ended up sufficiently Earthlike to potentially harbor reasonably familiar forms of life, or—indeed—to host and sustain us if we ever manage to reach them. If we could confirm life on other planets, then this would help answer deep existential questions about the uniqueness of life on Earth. But when exploring something as vast as space, it is important to hedge your bets based on a sound understanding of what creates the conditions you're searching for, to help focus firmly on only the most promising planets.

The importance of salt in the oceans is less grand. But the salt is very important nonetheless. Salt, together with temperature, determines circulation in the interior of the oceans. Thus, by understanding why and how the sea became salty, we can begin to think about the role of ocean circulation in shaping living conditions throughout the marine environment.

For a long time, it was thought that earliest Earth was so hot that no liquid water existed, and that all light elements were rapidly stripped from the inner solar system by the solar wind. If this were true, then the elements needed to form water on Earth would not have been freely available. As a consequence, it was proposed that a late bombardment by icy comets or similar gas- and water-rich materials brought the water to Earth after the planet had sufficiently cooled to retain it. This concept was supported by comparisons between the gas compositions of meteorites and Earth mantle rocks, notably using the noble gases krypton and xenon that do not react with other materials. There certainly is enough ice in space to have supplied our water (and atmosphere) in this manner.

And then the plot thickened. In July 2015, the probe Philae, which the European Space Agency's Rosetta mission landed on comet 67P/Churyumov-Gerasimenko (or "Churi"), discovered not only ice and dust, but also 16 types of organic compounds, present not in a loose distribution but in discrete clumps. Suddenly, the idea gained lots of traction that comets brought not only water, but also the ingredients for life, even in ready-made clumps! Intriguingly, in October 2015 it was reported that—as this comet slowly thaws—molecular oxygen (O_2) escapes in a constant and high (1% to 10%) proportion relative to water, which suggests that the comet also contains a surprising amount of primordial oxygen, which was incorporated during the comet's formation.

Other work favors an alternative explanation. This work found that the hydrogen isotope ratio of ice in comets may be different from that of water on Earth (isotopes of any element differ in the number of neutrons in their atom cores; we measure changes in the abundance ratio of these isotopes relative to an international standard, using mass spectrometers). It instead emphasizes that the chemical composition of water on Earth resembles that of the small percentage of water contained within rocky meteorites, and thus in asteroids, which essentially are very large meteorites. Thus, a theory was developed that the asteroids, planetesimals, and protoplanets that clumped together to form Earth had carried enough water in their rock minerals to explain our oceans. It would have escaped from the planet's interior as steam, which in turn would have condensed into water at the surface and in the early atmosphere. Calculations indicate that this mechanism can also provide plenty of water to explain Earth's observed water content.

The origin of Earth's water, and much of its atmosphere, therefore remains to be resolved. As more and more comets and meteorites are investigated in detail, we learn fundamental new things about the birth of our solar system and the potential routes by which key elements and compounds accumulated on the planets, including Earth. Planetary and space sciences are intimately interwoven with the Earth sciences.

We have a more complete understanding of the origin of salt in our oceans. It represents an accumulation of dissolved minerals over tens of millions to hundreds of millions of years. These minerals were broken up and dissolved during chemical weathering. We are all familiar with this process from limestone buildings that become pitted or smoothed by the action of water, wind, and weather; this is where the term weathering comes from. The key process at work is one of chemical reactions between the rock and the water, with an important role for gases that are dissolved in the water, such as carbon dioxide or sulfur dioxide, since these make the water more corrosive. The chemical weathering reactions break up rock minerals into charged atoms or molecules, called ions, which are removed in solution by river water and groundwater. This is exactly what happens when you dissolve table salt in water: the mineral salt breaks down into sodium and chloride ions that are held in a solution.

The early atmosphere contained high levels of carbon dioxide, or CO_2. This gas is easily dissolved in water, forming a mildly acid solution. In the CO_2-rich early atmosphere, this resulted in a corrosive, acid rain that was highly effective at chemically weathering rocks, and fresh volcanic rocks are especially easily weathered. The intense weathering released dissolved minerals in the form of ions into river water and groundwater. From early times onward, river and groundwater flow has transported the dissolved minerals to their final collection point, the ocean basins.

There is enough salt in the modern oceans to form a layer roughly 50 meters thick around the entire Earth's surface. Its buildup is a great illustration of the power of the geologic depth of time. To supply the amount of salt found in today's oceans, several hundred million years of river and groundwater flow would be needed, or several tens of millions of years of underwater volcanic and hydrothermal vent activity. To remove it would require many hundreds of millions of years of reactions with seafloor rocks and formation of minerals that get buried on the seafloor. The cycle of replacement of salts in the ocean is extremely slow—in technical terms, we say that salts in the oceans have very long residence times.

Given the extremely slow input and removal of salts, it becomes clear that the oceans' vast store of salt has accumulated because the oceans have for eons been the "end station" for salt transport. The weathering cycle has been feeding salts into the oceans over billions of years. Meanwhile, water itself continually evaporates from the oceans—concentrating its salts—and the evaporated freshwater then again cycles through the atmosphere to form rain, rivers, and groundwater, which eventually transport more salt toward the oceans. In similar ways, a terminal lake, which has only inlets but no outlet, will gradually accumulate salts. The only difference is that the world ocean has been at it for so much longer, and therefore has built up much more salt.

The ocean salt content built up early in the oceans' life cycle. Over time, the inputs and outputs of salts became closely balanced, so that the salt content of the oceans had stabilized at around its present composition and quantity many hundreds of millions or even well over a billion years ago. Interactions between the different dissolved ions that make up ocean salt are such that their relative proportions do not change much any more. So ocean water becomes more salty if water is evaporated off, or less salty if freshwater is added, but the average composition of the salts within it hardly changes.

As we have seen, both temperature and oceanic salt content, or salinity, were established very early in Earth's history. Spatial gradients in these properties are essential for movement of the atmosphere (temperature) and of the oceans (temperature and salinity). In consequence, we may safely infer that some form of atmospheric and ocean circulation existed from the earliest of times. And circulation is key for moving around heat, the ingredients of life, oxygen, and so on. Without circulation, nothing much would happen.

Spatial temperature gradients are unavoidable over Earth's surface because its spherical shape causes incoming solar radiation to be stronger at low latitudes than at high latitudes, because cloud patterns of whatever nature would always have imposed spatial gradients in reflectivity, and because land and sea surfaces have very different heat capacities. In consequence, there will always have been temperature contrasts on the surface of our planet. Temperature contrasts set up airflows that drive atmospheric heat transport, in a bid to compensate for net warming in some regions, and net cooling in other regions. Airflow means wind, and wind imposes drag on surface water. This sets in motion a surface circulation system of currents, small and large turbulent eddies, and enormous spinning current systems, or gyres, which can span the entire width of ocean basins.

The large-scale atmospheric-surface pressure systems, which result from low-latitude warming and polar cooling on a spinning globe, drive the major semipermanent or permanent wind systems. Key examples are the westward-blowing low-latitude trade winds, the eastward-blowing midlatitude westerlies, and the westward-blowing polar easterlies. To some extent, with varying intensities and latitudinal ranges, these systems will have been present at all times. Regional wind patterns result from interactions of these systems with the distribution of land and sea, which changes through geologic time. In addition, there are multiyear to decadal ocean-atmosphere interaction cycles, such as the El Niño Southern Oscillation, the North Atlantic Oscillation, the Arctic Oscillation, the Pacific Decadal Oscillation, and the Antarctic Oscillation. And finally, on even shorter time scales, we have the seasonal monsoons, and of course all the storms and other instabilities that make up the weather. All of these systems affect surface wind patterns and intensities, and thus surface-ocean circulation.

Long-term average surface-ocean circulation is dominated by long-term persistent wind systems. For instance, the combined action of the trade winds and midlatitude westerlies maintains big oceanic gyres between the equator and about 40 to 50 degrees of latitude in all ocean basins, which spin clockwise on the northern hemisphere and anticlockwise on the southern hemisphere. Earth's rotation squashes these gyres up against the western margins of the ocean basins, where they give rise to particularly powerful western boundary currents, such as the Gulf Stream in the North Atlantic, and the Kuroshio Current in the North Pacific. Another example of a long-term persistent wind system concerns the powerful and uninterrupted (by land) midlatitude westerlies of the southern hemisphere—known as the Roaring Forties and Furious Fifties. These drive the most powerful current of all: the Antarctic Circumpolar Current (ACC), which itself also wraps around the globe uninterrupted by land. Surface circulation generally affects only the upper few hundred meters of the oceans, but more than 30 million years of unimpeded driving by an amazingly persistent and intense wind system has allowed the ACC to extend from surface to seafloor throughout the Southern Ocean. The ACC truly is the great mixer of the world ocean.

Large-scale surface-ocean gyres likely existed as long as there have been oceans, since their driving wind systems—trades, westerlies, and polar easterlies—will have existed in some form and intensity in response to low-to-high-latitude temperature gradients on a rotating planet globe. In turn, western boundary currents will have existed whenever gyres were developed, as the planet's rotation would always squash gyres against western oceanic margins. Thus, reasonable expectations can be formulated for past large-scale surface-circulation patterns, based on estimates of the planet's surface-temperature gradients and past land-sea distribution, including any major ocean passageways between basins (figure 6). But in the discussion of ocean history, we are even more interested in deep-sea circulation, as it affects some 90% of the world ocean's volume, dominates property distributions through the oceans, and has been remarkably variable through time. Deep-sea circulation depends on subtle, relatively small-scale processes and contrasts between water masses, and therefore is considerably harder to estimate.

Deep-sea circulation is driven by seawater density differences. New deep water is formed from surface water in areas where temperature and salinity conditions cause surface densities that are similarly high to those at depth. This allows vertical mixing of the water column. Continuous supply of dense water from the surface sets up a flow of water down to the depth where it belongs according to its density in relation to the vertical density profile in the region. Then it spreads sideways, displacing less dense ambient water upward. The ambient water had become less dense because of turbulent mixing, which almost always develops as water moves, and especially as newly formed deep water sinks through the water column. Mixing is also amplified over rough seafloor topography. But what actually determines seawater density in the first place?

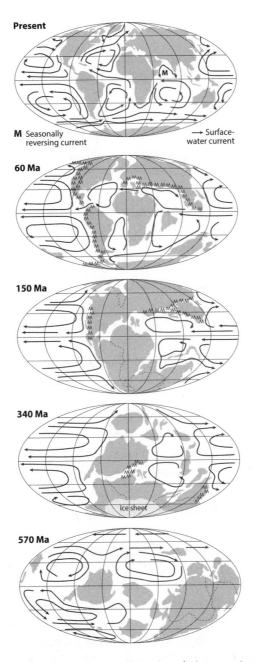

Figure 6. Wind-driven surface-circulation patterns in today's oceans (*top*), and—in highly schematic form—at selected past time slices (*lower panels*). In the present-day panel, M in the northern Indian Ocean marks a current that reverses seasonally in response to Asian monsoon changes. Ancient surface circulations have not been recorded by measurable properties, and thus are shown for illustration only using considerable artistic license (along with any modeling information), based on general principles of wind-driven circulation in interaction with continental distributions. Ma stands for millions of years ago.

Seawater density depends on three things: pressure, temperature, and salinity. Pressure in the oceans is just a function of depth. As a rule of thumb, pressure goes up by about 1 atmosphere for every 10 meters of depth. So, at the very surface, pressure is 1 atmosphere (the atmospheric pressure), at 10 meters' depth it is about 2 atmospheres, at 20 meters' depth it is about 3 atmospheres, and so on. The increase is not exactly 1 atmosphere per 10 meters, but just a bit less. As a result, pressure at 4000 meters' depth is closer to 398 atmospheres. The very high pressures in the deep sea cause the water to be compressed a little. This implies a reduction in volume for the same mass, and thus an increase in density (mass per volume). But water is pretty amazing because it is not actually that compressible at all. Even at 4000 meters' depth, seawater is compressed by less than 2%. In consequence, the influence of pressure on density is very small in the oceans. I will ignore it here, as it's not needed for our further discussions.

This leaves us with the two main influences on seawater density: temperature and salinity. As a result of the dissolved salts, seawater freezes at a lower temperature than freshwater—namely, at around –1.9°C. Like freshwater, seawater expands as temperature increases. Freshwater is a bit special in that it reaches its highest density at 4°C, even though the freezing point is lower at 0°C, which is why the deep-water temperature in deep lakes of cold regions is 4°C. Because of the salt it contains, seawater shows none of that funny behavior. Instead, seawater reaches its highest density at its freezing temperature of about –1.9°C, and then steadily expands, and thus becomes less dense, as temperature goes up. This temperature dependence is technically referred to as "thermal expansion." Today, ocean temperatures vary over a wide global range between the freezing point and more than 30°C, and all the high values are restricted to the upper few hundred meters; deep-sea temperatures range between about 0°C and 3°C, around a global average deep-sea value of about 2°C. Yes, believe me, the deep sea really is that cold, even today in a relatively warm interlude between ice ages. I have had many painful reminders of it during research cruises, when we were rinsing and overfilling sample bottles on deck from the containers that had brought the water up—a set of those swiftly reduces your fingers to numb icicles. During the last ice age, about 25,000 to 19,000 years ago, deep-sea waters were even closer to the freezing point, so it's just as well that our ancestors weren't tempted to go sampling them. We will see that, during really warm episodes in Earth's history, like the time of the dinosaurs, deep-sea temperatures did manage to rise to 15°C or even 20°C. This represents a remarkably major amount of excess heat storage relative to the present.

Ocean temperature increases by absorption of incoming solar radiation. It decreases by emission of long-wave radiation, loss of so-called "latent heat" upon evaporation, and transfer of heat by conduction and convection. But let's not get distracted by what these various terms exactly mean. The important bit is that heat exchange happens almost exclusively at the ocean surface. There is a very small geothermal heat component through the ocean crust; it's tiny compared with the other terms. As a result, temperature behaves conservatively in the interior of the ocean, which means that it changes only as a result of circulation and mixing with other water masses. Surface temperature is high in low latitudes, and gets lower toward high latitudes. In our example of the time of the dinosaurs with really warm deep-sea conditions, therefore, it must have taken a long time for the heat to enter the deep sea from the surface, through circulation and mixing. Also, we can infer that the coldest surface areas must have been at least of the same temperatures as the deep sea at that time.

Salinity is the other property that affects sea-water density. On average, the proportion of dissolved salts in seawater is close to 35 grams per kilogram. Modern measurements of salinity are made using conductivity, relative to an international standard. Because they are made in relative terms, such measurements no longer have units to mention. So we'd simply say that "the salinity is 35." It may feel a bit odd at first, but you'll soon get used to it. As the salinity goes up, seawater density goes up as well; in technical terms, we speak about "saline contraction." The modern global salinity of seawater mostly ranges between 33 and 37, but in some places like the Red Sea it can go up to 40. Also, it can drop to well below 30 in exceptional places with a lot of freshwater input. Completely pure (distilled) freshwater has a salinity of zero.

Seawater salinity goes up because of freshwater removal by evaporation, and goes down because of freshwater addition. Both these processes are limited to the sea surface and depend on the atmospheric freshwater cycle. Surface-water salinity is high in evaporative subtropical regions, and lower in rainy equatorial regions and especially in low-evaporative and rainy mid- to high-latitude regions. Within the interior of the oceans, however, salinity can change only through mixing with other water masses; like temperature, salinity behaves conservatively in the interior of the oceans. In consequence, salinity is an excellent property for tracing water masses from one place to the next.

The influences of temperature and salinity on seawater density depend on each other. When considering the impacts on deep-sea circulation, we can focus on an appropriate deep-sea temperature range for most of Earth's history. During the last ice age, a cold interlude of the current icehouse period, average deep-sea temperature was about 0°C. Today, average deep-sea temperature is close to 2°C. We will see that deep-sea temperature was about 8°C at around 60 million years ago, increasing to about 12°C at around 50 million years ago. During much of the warm Mesozoic greenhouse period of the dinosaurs, deep-sea temperature seems to have been 10°C to 12°C as well, with highest reported values for any time of up to 20°C. In short, the temperature window over which we need to consider temperature and salinity impacts on density spans a range from a minimum of –1.9°C to a maximum of 20°C. Within this deep-sea-temperature window, a given salinity change has more impact on density at low temperatures than at high temperatures, while a given temperature change has less impact on density at low salinity than at high salinity. Given that global average temperature has changed considerably over Earth's history, the salinity sensitivity change at different background temperatures is the most important here. Deep-sea "background" temperatures responded to those changes within a few thousand years, as a result of deep-ocean circulation.

For a mental image of how temperature and salinity can drive deep-sea circulation, consider our modern situation. Today, we still are unmistakably living in an icehouse world, with vast ice coverage that varies greatly over time scales of tens of thousands to hundreds of thousands of years. The most recent ice age may have ended about 19,000 years ago, but Antarctica, the continent over the south pole, still is completely covered in ice. In the high latitudes of the northern hemisphere, there is another major continental ice sheet (Greenland), along with many smaller ice caps such as those in Alaska. The Arctic Ocean contains a large sea-ice cover, part of which—for the time being—still survives through summer. The average temperature of the entire world ocean is only 3.5°C, because it is dominated for more than 90% by the massive deep sea, which is only about 2°C. We are living in a very cold world, which is clear for all to see from the geologic record; it's a low-CO_2 world, with a natural level of 280 ppm for warm interludes like the current one and a level of 180 ppm during ice ages. Of course, we're changing this very rapidly now, through human-induced emissions, and have already reached 400 ppm.

During icehouse conditions, the low latitudes are a bit cooler than during warm greenhouse climate states, such as the Cretaceous period of about 145 to 66 million years ago. Most notably, however, the high-latitude polar regions are very cold because of the ice and snow cover at high latitudes, which reflects a lot of incoming solar radiation. As a result, very strong temperature gradients exist in the oceans and over land. With persistent freezing conditions, high latitudes are the key regions of high surface-water density. This is especially true at the margins of sea-ice cover, where brine rejection makes the water more saline, and in places like the North Atlantic where high-salinity subtropical water is brought poleward by the Gulf Stream. These salinity additions in cold regions create optimum high-density conditions.

High-density surface water sinks down into the interior of the oceans, setting up a vigorous deep-water circulation. The return upward flow is achieved by turbulent mixing within the interior of the oceans, as mentioned above. The first to advance a concept of cold deep currents that originate from the poles was Benjamin Thompson, Count Rumford, in the late 1790s. But it was not until the late 1920s that the Austrian oceanographer Albert Defant introduced the modern concept, which includes the roles of both temperature and salinity, and which he accordingly named the "thermohaline circulation" (figure 7).

Water formed in the high-latitude North Atlantic is relatively saline owing to the subtropical influence. Strong cooling transforms

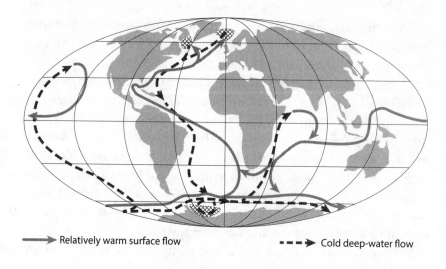

Relatively warm surface flow   ▪▪▪▪▶ Cold deep-water flow

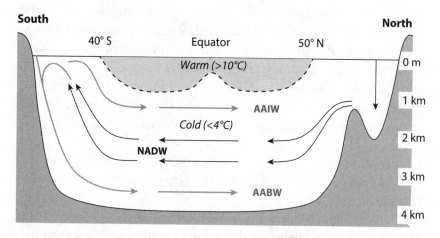

Figure 7. Schematic representation of the interconnected "conveyor-belt" circulation in to-day's world ocean (*top panel*), with vertical cross-section through the Atlantic Ocean that shows the main water masses (*bottom panel*). In the top panel, deep currents (*dark gray, dashed arrows*) emanate from regions of deep-water formation (*hatched*), and eventually reconnect with returning surface currents (*light gray arrows*). AABW, Antarctic Bottom Water; AAIW, Antarctic Intermediate Water; NADW, North Atlantic Deep Water.

it into a water mass called the North Atlantic Deep Water (NADW), which spreads southward throughout the Atlantic at depths of several hundred to about 3500 meters. Below the NADW and down to the greatest depths, the Atlantic is filled with Antarctic Bottom Water (AABW), which originates from processes associated with freezing in the sea-ice-covered Weddell Sea (see figure 7). Sea ice has a very low salinity, which means that salt is not included when seawater freezes to form sea ice. The salt is rejected as high-salinity brine into the remaining seawater, so that surface-water salinity goes up around areas with active sea-ice formation. This water is also very cold—almost frozen. The combination of injected salt brine and very low temperatures makes regions of sea-ice formation prime candidates for new deep-water formation. The AABW spills over into the Southern Ocean, and then spreads northward in the Atlantic at great depth. Above the NADW, another water mass from the high southern latitudes extends northward, up to about 20° N. This is the less saline but cold Antarctic Intermediate Water (AAIW), at depths of about 700 to 1200 meters.

Intense mixing in the violent Southern Ocean around Antarctica creates a single water mass called Circumpolar Deep Water (CDW) from combination of NADW and AABW. This CDW flows northward into the Indian and Pacific Oceans, filling them from the greatest depths up to 1200 meters or so. Above that, between roughly 700 and 1200 meters, the less saline but cold AAIW also spreads northward into these oceans.

There are two minor regions of deep-water formation in the modern ocean that are strikingly different from the dominant high-latitude regions. These are the Mediterranean Sea, and the combined Red Sea and Persian Gulf. Deep waters formed in those arid regions have high salinities of 38 to 40 and are relatively warm, at about 13°C for the Mediterranean and 23°C for the Red Sea. Outflow from the Mediterranean is only 2°C or 3°C colder than inflow from the Atlantic, and for the Red Sea outflow this difference is only 0.5°C. This indicates that deep-water formation in these basins, located in warm and arid subtropical regions, depends more on salinity increase than on cooling. As such, the processes of deep-water formation in these basins have often been considered similar to the warm deep-water formation of past greenhouse periods, when no strong temperature gradients existed. We will have a look at this when our discussion gets to the Mesozoic.

As new deep waters enter into the interior of the oceans, they displace ambient, less dense waters, as mentioned above. These displacements eventually come close enough to the surface that they become entrained into the wind-driven surface circulation. Surface circulation is connected between ocean basins through openings such as—for example—the sea straits through the Indonesian archipelago, the passage around Cape Horn, Drake's Passage, and the Bering Strait. By means of the surface circulation, the entrained former deep waters from the ocean interior are returned toward regions of deep-water formation, closing the loop that is sometimes referred to as the "conveyor-belt circulation" (see figure 7, upper panel).

Deep-water circulation is how the deep sea breathes. It is the only means of transporting oxygen into the deep sea. The oxygen enters surface water through gas exchange (or equilibration) with the atmosphere, which is dependent on atmospheric oxygen concentrations, water temperature, and—to a lesser extent—water salinity. In regions of deep-water formation, it is then carried from the surface into the deep sea within the newly formed deep water. Oxygen is more soluble in water at low temperatures than at high temperatures; from 0°C to 30°C, the oxygen solubility drops by almost 50%. For a given atmospheric oxygen concentration, this means that cold water can hold more oxygen than warm water. As a result, mechanisms of cold deep-water formation are more efficient at oxygenating the deep sea than mechanisms of warm deep-water formation.

All major deep-water masses in today's oceans are cold. They were efficiently formed in large volumes from cold high-latitude surface waters that absorbed a lot of oxygen through gas exchange with the atmosphere. Thus, the modern ocean has a vigorous deep-sea circulation, involving water masses that—at source—start off with high oxygen concentrations. This results in a modern deep ocean that is, overall, well oxygenated. As oxygen is critical to the metabolic processes by which animals oxidize, or burn, their organic food matter to obtain energy, high oxygen concentrations offer relatively easy living conditions for most, except for low-oxygen specialists. Waters that contain no oxygen (anoxic), or even free hydrogen sulfide (euxinic), are lethal to most animals. This is clearly illustrated in so-called "dead zones," where runoff of fertilizer, wastewater, or sewage into coastal settings or lakes creates massive algal blooms, whose decomposition depletes all the dissolved oxygen. Under such conditions, massive mortality occurs throughout the ecosystem. Most of us will have seen the devastation of anoxic conditions in ditches, ponds, or even in a fish tank if that broken air pump was ignored for too long.

LIFE, OXYGEN, AND CARBON

The origin of life from chemical compounds is a topic of great debate. At some stage, three major classes of biomolecules need to have arrived at the scene: amino acids, nucleic acids, and lipids. These are essential for the earliest forms of life to start up. The compounds may have been carried onto Earth by comets, as suggested by the recent findings of the European Space Agency's Rosetta mission. Alternatively, these compounds may have formed on Earth from reactions that involve hydrogen cyanide, hydrogen sulfide, and ultraviolet light. The hydrogen cyanide may have been imported by comets, which contain quite a bit of it, or may have formed from reactions between hydrogen, carbon, and nitrogen that were triggered by the energy of comet impacts. Hydrogen sulfide was abundantly present on early Earth, and so was ultraviolet light. We won't go into this hotly debated topic any further, and instead jump to the point where life in a cellular shape appeared.

Tracing the beginning of cellular life is a game of molecular cat and mouse. Finding signs of early life begins with finding the oldest rocks on Earth—not an easy feat in itself. Back then, life was single celled, on the scale of bacteria. There are billions upon billions of bacteria around us and inside us every moment of our lives, but how many have you actually ever seen? With that in mind, imagine going to some extremely ancient rocks and trying to find fossilized remains of these tiniest of organisms. If no organic matter is preserved, then you can forget it immediately. If it is preserved at all, then the vast bulk of it will be degraded and shapeless after billions of years. But over time, researchers have been finding more and more evidence, and have been gaining more and more detailed insights. The search for early life has turned into a search for the oldest strategies used to make organic matter from inorganic compounds: synthesis. This requires energy, which can be had from sunlight, leading to photosynthesis, or from chemical reactions, leading to chemosynthesis. Either way, it's widely accepted that it happened in watery environments.

When we go looking for the functional equivalent of photosynthesizing land plants in today's oceans, our eye is drawn to algae. As a technical aside, note that genetic research has classed some marine algae among the plants under the supergroup Archaeplastida, and most of them in another supergroup called Chromalveolata. In brief, the genetic studies found that life is organized in much more complex lineages than the traditional subdivisions of animals, plants, fungi, and protists that used to be taught in times before genetic research got involved (including the times that I went to university; at which time we used the name protists for all unicellular eukaryotes—organisms with cells that contain a nucleus and other organelles, enclosed within membranes). Algae produce organic tissue via the process of photosynthesis. But when we look in more detail, it turns out that the real masters of this game are not the algae themselves, but bacteria. The chloroplasts that do the job of photosynthesis within multicellular organisms such as algae and plants started out as symbiontic photosynthesizing bacteria called cyanobacteria. Yet life did not likely start with photosynthesis, as photosynthesis requires complex reactions that in turn require complex organic compounds.

We saw previously that the early Earth's atmosphere and oceans contained no free oxygen. So, early life must have developed in the absence of free oxygen, which means that there are two possible pathways, either chemosynthesis or non-oxygen-producing "anoxygenic" photosynthesis. This makes sense because no living matter can form spontaneously from inorganic compounds in a world with free oxygen, where organic matter is too rapidly oxidized (it decays).

For some time, the very first chemical traces of life were known from graphite in sedimentary rocks in Greenland, dating back to 3.7 billion years ago (see figure 1). The first known actual fossils of life were microbial mats dating back to about 3.48 billion years ago, which were discovered in the Pilbara region of Western Australia. And by 3.43 billion years ago, more complex microbial mat structures, called stromatolites, formed a shallow-water reef-type formation that was also found in the Pilbara region. Very recently, these findings were trumped by a new finding of stromatolite fossils in southwestern Greenland that date back to 3.7 billion years ago. Similar stromatolite structures still exist today, most notably in Shark Bay, Western Australia. These mounds comprise layers of microbial mats of cyanobacteria, or blue-green algae, in shallow waters where sunlight provides the energy needed to synthesize organic tissue; these organisms are photosynthesizing. But that does not mean that the organisms that built up the ancient stromatolites necessarily produced oxygen. The production of oxygen in photosynthesis relies on the "oxygenic" pathway of photosynthesis. Instead, the most ancient type of photosynthesis known is that of purple sulfur bacteria, which follow the anoxygenic pathway that produces hydrogen sulfide (H_2S; infamous for its rotten-eggs smell) rather than oxygen. Today, both pathways are known to be active in different levels of stromatolite mounds, and some species of cyanobacteria can actually photosynthesize using either pathway.

So, might anoxygenic photosynthesis be the first way in which life managed to synthesize organic tissue? Not likely. Chemosynthesis is still the way most researchers consider to be the most ancient. A common train of thought is that the last universal common ancestor (LUCA) of all three major groups of life on Earth—the Bacteria, Archaea, and Eukarya—was similar to the types of high-temperature-resistant microbes seen on hydrothermal vents. In those pitch-dark deep-sea environments, chemosynthesis is the key process.

A complicating twist to the tale arose when it was discovered that tolerance to high temperatures is a more recent adaptation, and therefore that LUCA more likely lived in temperatures below about 50°C. That opens up the possibility that it lived at or near the surface, with access to light, chalking up one point in favor of potential photosynthesis. But then, chemosynthetic microbial communities have been found living inside oceanic crust at much lower temperatures than those seen at vents. For example, an extensive community has been found at a location with temperatures around 65°C, at 350 to 580 meters' depth inside ocean crust, which relies on chemical reactions between iron-rich basaltic crust material and seawater that enters the crust through cracks. And chemosynthesis is now also well known from many other relatively cool environments. So nothing is ruled out yet; the search is still on. For now, chemosynthesis remains firmly in the lead as the favored process for early life.

After LUCA, life split into three primary lines of descent, but there remains much debate about exactly how this happened. Regardless, we ended up with one line that led to the Bacteria, including photosynthetic bacteria, and another that led to the Archaea. These all consist of small, simple cells without a nucleus or other organelles, which are known as prokaryotes. The third lineage gave rise to the Eukarya. Eukarya have cells of the previously mentioned eukaryotic type, which are larger and have a nucleus and organelles such as the mitochondria. The Eukarya eventually developed into a very rich diversity that includes all higher multicellular plants and animals, as well as many single-celled organisms such as slime molds, flagellates, amoebae, and—critical in today's ocean plankton—a wide variety of photosynthesizing algae and both grazing and predating/scavenging foraminifera.

It is thought that the primary splits into Bacteria, Archaea, and Eukarya may pre-date even the oldest fossil stromatolites, in which case all three main lineages are extremely ancient. What is not known is when the early Eukarya began to develop what we now consider to be the typical eukaryotic cell type. But fossils of the tube-shaped *Grypania* from Michigan indicate that this happened nearly two billion years ago.

At some stage, photosynthetic bacteria entered eukaryotic cells, probably by being ingested for food in the first place, and survived. Surviving inside the larger cells, their photosynthesis provided food to both themselves and the host cells, and eventually the photosynthetic bacteria replicated inside the hosts. The replication of such bacteria occurs by asexual cell division, and therefore is simpler and easier to continue than sexual reproduction would be. A stable symbiontic relationship developed.

Not long before one billion years ago, the eukaryotes had split into a line that had become capable of photosynthesis, and that started the lineages of plants and algae, and other lineages including those leading toward fungi and animals. Multicellular algae, or colonies of unicellular algae, first appeared before 1 billion years ago; recent findings in China date back to 1.56 billion years ago. Animals, by definition multicellular, appeared at around 0.9 billion years ago. The period from roughly 2 to 1 billion years ago is sometimes referred to as the "boring billion" for its lack of obvious evolutionary changes.

And then, finally, after some four billion years of Earth history, we get to the appearance of macroscopic animals. As far as we know from the fossil record—and we may well be missing information because it has not been preserved or it has not been found yet—there were at least two explosions of macroscopic life, when surprisingly diverse biota burst onto the scene (see figure 1). First came the Avalon explosion of the Ediacaran biota between 575 and 542 million years ago. The Ediacaran biota went extinct toward the end of this period. Then followed the more famous Cambrian explosion between 542 and 520 million years ago. All this was taking place in the seas, or at least in water.

The first primitive plants appeared on land, along with fungi, roughly 450 million years ago. Intriguingly, though, lobster-sized centipede marine animals seem to have at least temporarily checked out "what's cooking" on land well before that, as early as 530 million years ago, leaving fossilized footprints. Experts conjecture that they probably went ashore only to mate and lay eggs, as do modern horseshoe crabs, or that they were trying to escape predators or scavenging for food. I guess they may not have liked it much, finding only bare rock and rivers and pools with bacterial slime and algae. In any case, it took another 30 million years before animals properly ventured onto land; scorpion-like animals were among the first to do so, at around 420 million years ago. The first vertebrate animal known to have occurred on land dates back to roughly 375 million years ago. It is known as *Tiktaalik*, a type of lobe-finned fish that represents an evolutionary transition between fish and amphibians.

We cannot go through the development of life without considering the history of oxygen and the cycling of carbon in the atmosphere and ocean, as these are all closely related. In the following, I will briefly outline the main points, and will add further details at relevant points along our journey through ocean history.

Oxygenic photosynthesis is more efficient than anoxygenic photosynthesis. The earliest stromatolites are thought to have consisted of cyanobacteria or similar organisms, initially using anoxygenic photosynthesis only. They appear to have added the new trick of oxygenic photosynthesis at around 2.8 or even 3.2 billion years ago. It is hard to pinpoint exactly when oxygen first started to be produced. The early atmosphere and ocean were full of reduced chemical compounds, meaning they existed in a nonoxidized state. As soon as some oxygen was produced, it would have been consumed directly in chemical reactions oxidizing these reduced compounds. This means that we would not see any significant signs of free oxygen, even if some of it were produced.

By about 2.5 billion years ago, oxygen production accelerated, which we recognize by means of the onset of the Great Oxygenation Event (see figure 1). Chemical data suggest that the first oxygen-breathing life (types of bacteria) on land may have appeared about 2.48 billion years ago, which would indicate that some free oxygen had become available in the atmosphere by that time. Other evidence comes from the formation of oxidized iron deposits on land, known as redbeds, and from a major increase in deposition of so-called "banded iron formations" (BIFs) between about 2.5 and 2.3 billion years ago. BIFs are underwater depositions of oxidized iron in truly gargantuan quantities—most iron mining today relies on these deposits, including the intense mining in Australia.

The deposition of BIFs continued for a long time after their peak onset, at least until 2 and possibly 1.8 billion years ago. Given that reduced iron is soluble in water, while oxidized iron is not, the deposition of BIFs is strong evidence that the oceans began to receive oxygen, but also that all the oxygen was drawn out straightaway by chemical oxidation reactions. Because of atmosphere-ocean gas exchange, this also kept atmospheric oxygen levels low, and the first convincing evidence for increasing oxygen levels comes from the appearance of complex multicellular life at around one billion years ago.

Life's impacts on carbon cycling start at the foundation of food webs with primary production, a technical term for the production of organic tissue via photosynthesis or chemosynthesis. The term food web is perhaps more familiar in its previous guise of food chain. But the flow of food from primary productivity upward is rarely in one single direction, as the word chain would suggest. The revised term of food web is a more appropriate reflection of the complexity of the actual links. In the modern oceans, photosynthesis is by far the predominant process for new organic-matter generation. The primary productivity from this process therefore feeds most of the world ocean's inhabitants, as well as birds and land animals that eat food from the ocean. Chemosynthesis cannot be ignored, but it has much less impact.

Let's have a look in a bit more detail. Sunlight penetrates from the surface down into the water column to a maximum of about

200 meters, but commonly much less than that. This illuminated layer, down to a depth where the light intensity is reduced to 1% of that at the surface, is called the photic layer. Organic tissue can be chemically summarized as a carbohydrate, or CH_2O. Primary production through photosynthesis can then be represented as: $CO_2 + H_2O \rightarrow CH_2O + O_2$, and the energy for this reaction is obtained from sunlight. Decomposition of organic matter, often referred to as respiration, follows the opposite direction: $CH_2O + O_2 \rightarrow CO_2 + H_2O$. The decomposition reaction releases energy, which is how you and I—and animals in general—obtain our energy: we eat organic matter, react it with oxygen that we breathe in (we oxidize the organic matter), gain the energy, and breathe out the CO_2. As for the water, you can probably guess. So, plants consume CO_2 and nutrients and produce organic matter and oxygen, while decomposition—either inside guts or outside them—consumes organic matter and oxygen and releases CO_2, nutrients, and energy.

The food produced in the oceans' photic layer doesn't just feed ecosystems of swimmers (the pelagics) in the upper few hundred meters, but—through sinking of dead matter to the deep sea and seafloor—also pelagic animals in the deep sea and animals living on and in the seafloor (the benthics). The flow of food energy and nutrients in the ocean therefore goes from the top to the seafloor. We call this the biological pump. Put plainly, the deep sea is fed by corpses and excrement.

Marine pelagic food webs can be remarkably short. In productive regions, such as places where deep water laden with nutrients is circulated or "upwelling" toward the surface where there is abundant sunlight, the links can run very fast from marine algae to zooplankton to the largest animals like whales. In such systems, there is such plentiful resupply of nutrients that the ecosystem can be quite wasteful, and a considerable portion of dead and decaying matter sinks out of the photic layer toward the seafloor. In most of the oceans, nutrient resupply is much weaker, and this favors a very complex network of links with highly efficient recycling of matter through lots of specialist organisms that occupy narrow ecological windows, or niches. In such regions, not much organic matter drops out of the photic layer.

On average, between 1% and 25% of the organic matter formed by photosynthesis drops out of the photic layer. Of this, 9 out of 10 parts

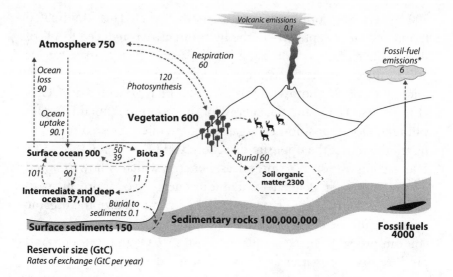

Figure 8. Schematic of key carbon reservoirs and rates of carbon exchange between them. The asterisk with fossil-fuel emission indicates an artificial net addition to the atmosphere-hydrosphere-biosphere system that is not further included in the natural equilibrium exchanges shown in the diagram (roughly 40% of it goes into the ocean, and 60% stays in the atmosphere). GtC is gigatonnes of carbon, where one gigatonne = one billion metric tons.

are recycled in the deep pelagic and benthic ecosystems, and only 1 out of 10 parts is buried in the seafloor sediments. So, on average, only something between 0.1% and 2.5% of organic matter produced in the photic layer is lost from the oceanic ecosystems by being buried away in the sediment. The proportion of buried organic matter is above average under the more wasteful upwelling regions, and below average under the more efficiently recycling regions where new nutrient supply is very low.

The proportion of organic matter that gets buried in the sediments represents a net loss of carbon from the ocean-atmosphere system. Given enough time—in the order of many thousands to millions of years—this loss can build up a large net carbon removal from that system, reducing CO_2 levels. That lost carbon only gets back into the atmosphere-ocean system when the carbon-containing sediments are uplifted and weathered, or when the ocean crust containing the sediments is subducted and melted in the Earth mantle, followed by volcanic outgassing (figure 8).

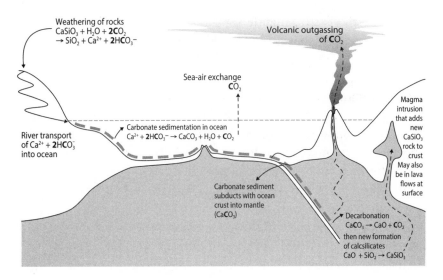

Weathering of rocks
$CaSiO_3 + H_2O + 2CO_2$
$\rightarrow SiO_2 + Ca^{2+} + 2HCO_3^-$

Volcanic outgassing
of CO_2

Sea-air exchange
CO_2

Magma
intrusion
that adds
new
$CaSiO_3$
rock to
crust
May also
be in lava
flows at
surface

River transport
of $Ca^{2+} + 2HCO_3^-$
into ocean

Carbonate sedimentation in ocean
$Ca^{2+} + 2HCO_3^- \rightarrow CaCO_3 + H_2O + CO_2$

Carbonate sediment
subducts with ocean
crust into mantle
($CaCO_3$)

Decarbonation
$CaCO_3 \rightarrow CaO + CO_2$
then new formation
of calcsilicates
$CaO + SiO_2 \rightarrow CaSiO_3$

Figure 9. Schematic representation of inorganic controls on carbon cycling associated with the long-term cycle of plate tectonics and weathering.

On time scales of thousands of years, the biological pump of carbon from surface waters also removes CO_2 from the atmosphere. But on these time scales, the sediments are not involved, and we're only looking at a net movement of carbon from the atmosphere and uppermost ocean into the deep sea. Overall, the deep sea holds a lot of carbon. Some of it is lost by interaction—over 10,000 years or more—with carbonate in the sediments. This process is called carbonate compensation, and it involves the inorganic (mineral) carbon cycle that revolves around the formation and burial of carbonate shells and skeletal parts, and their dissolution at other times.

The organic side of the global carbon cycle is immediately obvious when we think a little about life and death in the oceans. The inorganic (mineral) carbon cycle is much less familiar. I for one had never heard much about it until my first introductions to Earth chemistry at university, and I don't think it features in any high-school or college curriculum even today. It's straightforward in general terms (figure 9). The key elements are weathering of rocks, which consumes CO_2, and burial of carbonates in ocean sediments (either as shell and skeletal components, or as chemical precipitates) that are eventually taken into the Earth's mantle via subduction, where they decarbon-ize to release CO_2 that is outgassed by volcanoes. In detail, however, the inorganic carbon cycle is complicated, especially in its interaction with the organic carbon cycle. I will introduce appropriate details where relevant to the various topics that will be discussed. In particular, this will be in the Reverberations section of chapter 4, and in chapter 5. But at this initial stage, we do need to mark a few key steps in the inorganic-carbon-cycle development through Earth history.

In the long time from the appearance of the first oceans (some four billion years ago) to about 542 million years ago, the oceans held very few to no organisms that formed carbonate skeletal parts. Carbonate precipitation in those times occurred abiotically—that is, purely chemically—with contributions from calcified cyanobacteria. In reference to their limited content of life, the oceans of these times are popularly referred to as carbonate Strangelove oceans, after a mad nuclear-holocaust expert of the same name in the 1964 Stanley Kubrick movie *Dr. Strangelove*. The name was first used in relation to an ocean in which a mass extinction had (nearly) wiped out all surface life.

From about 542 million years ago, organisms appeared that were capable of building carbonate skeletal parts. These came to domi-nate the seafloor, especially in shallow-water regions. No open-ocean carbonate-shelled plankton existed yet, which are technically known as planktonic calcifiers. In reference to the scientific name of "neritic zone" for shallow-water regions, the oceans of this time with pre-dominant biological carbonate deposition in shallow regions have been named Neritan oceans. Neritan ocean conditions prevailed un-til a major mass extinction at 252.3 million years ago.

After the extinction event of 252.3 million years ago, planktonic calcifiers started to develop (see figure 2). Over about 100 million years, they developed into a key driver of the inorganic carbon cycle. They have remained so ever since, until today. In reference to the Latin word *creta*, which means chalk, the oceans with planktonic cal-cifiers have been named Cretan oceans.

The shuffling of carbon between atmosphere and deep sea causes CO_2 fluctuations, of the order of thousands to a hundred thousand years, that are important for climate. But as soon as the sedimentary inorganic carbon cycle gets involved too, there can be net loss that leads to longer-term underlying fluctuations over time scales that span many millions of years. Thus, life's interactions with different levels of the ocean and its sediments have caused greenhouse-gas variations over a wide variety of time scales, all of which are relevant to climate. And these interactions were of different natures and intensities in Strangelove (4 billion to 542 million years ago), Neritan (542 to 252.3 million years ago), and Cretan oceans (252.3 million years ago to present), because of their fundamental differences in the processing of carbon.

This brings us to a point where we need to more systematically review the controls on climate, including its links with changes in the world ocean and its internal geologic, chemical, and biological processes. These interactions between climate and the oceans represent a complex web of processes that work in both directions: from climate to the oceans, and from the oceans to climate. As we go through these controls on climate change, we will encounter many examples of the complex mutual influences and interactions. Once that context has been digested, we are ready to launch into our main case histories of critical events in ocean history.

CHAPTER 3

CONTROLS ON CHANGE

When aiming to discuss ocean changes, the topic of this book, we have to address climate changes because the oceans and climate are intimately entwined. We simply could not consider climate variations on Earth without scrutinizing their ultimate cause—no matter what period of geologic history—which is the balance between energy received by Earth and energy lost (to space). It really is as simple as this: when more energy comes in than is lost, Earth warms up, and when more energy is lost than received, Earth cools down. The energy balance is measured at the top of the atmosphere, the boundary between Earth and space.

In the discussion of climate change, including what's happening today, the energy balance is all-important, yet many people never think about the problem in those terms. We hear lots of arguments about natural variability and how our emissions cannot be important, but the truth is that talking about global climate change can be easily reduced to arguments of energy in versus energy out, on a planetary scale. When that is done, natural variability can be understood, and the direction of change for the future becomes inevitable. If the Sun hardly changes (as is the case), and the well-understood retention of heat due to greenhouse-gas concentrations does change (as is the case), then there will be constant energy coming in and a reduction of energy going out—this will lead to warming, in the same way that eating more calories than you burn will make you gain weight. So let's have a look at what's involved.

The Sun is the grand master of incoming energy. Today, at the top of the atmosphere, Earth continuously receives about 340 watts per square meter, averaged over the entire global surface area. About 30% of this is reflected back into space by clouds and from the surface; we call this the albedo effect. The remaining 70% of the 340 watts per square meter, or 240 watts per square meter, are absorbed by the surface, which causes warming. For scale, this equates to a sizable (1500-watt) room heater going full blast, day and night, on each and every patch of 2.5 by 2.5 meters of Earth's surface. Given the intensity of the incoming energy, important impacts on climate could result from even small changes in the albedo, the reflectivity percentage.

Note that we will ignore any surface warming due to heat escape from the hot interior of Earth (including volcanoes) because—at only 0.09 watts per square meter—this today is almost 3000 times smaller than the net energy received from the Sun. In the earliest phase of Earth, heat escape from the interior was higher than today, as Earth was warmer and the concentrations of elements undergoing radioactive decay were greater. Model calculations suggest that at around 4.6 to 4.5 billion years ago the interior heat escape may have been some four times greater than today, roughly 3 billion years ago it may still have been two times greater than today, and 500 million years ago it would have approximated present-day values within about 25%. It is clear, therefore, that heat escape from the interior has on average never been as significant to climate as the influence of the Sun, even if we allow that the Sun started at about 70% of its modern intensity and gradually intensified until the present.

The total amount of solar radiation received and its distribution over the Earth's surface are not constant through time. Changes in the Earth's orbital configuration around the Sun have important impacts on climate and are introduced first. Next, we will consider the Sun's output itself, which is not exactly constant either.

Greenhouse gases are the next item on our list because incoming energy is only one side of the energy balance. The other concerns the efficiency at which Earth cools down. As any body that gets warmed up, Earth cools by losing heat through emission of thermal infrared radiation. But some of that outgoing radiation gets blocked in the

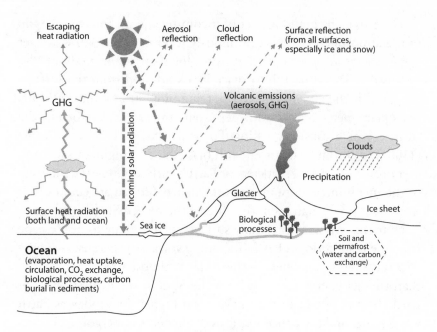

Figure 10. Main components of Earth's climate system. GHG stands for greenhouse gases.

atmosphere by greenhouse gases, which act a bit like an insulating blanket that causes warming of the surface.

Apart from regulating greenhouse-gas concentrations, plate tectonics also affect the distribution of land versus sea, and the location and intensity of mountain-range development. Finally, there are occasional impacts in the purest sense of the word; namely impacts of bodies from space, such as meteorites, asteroids, and comets.

Before we jump in at the deep end, we need to identify the basic component parts of the Earth's climate system, and their fundamental interactions (figure 10).

The climate system comprises a lot more than just the atmosphere. The watery component of Earth's surface, which we call the hydrosphere, directly interacts with the atmosphere. The oceans are by far the largest part of the hydrosphere and are very important in the climate system because of their great capacity to store, move, and release massive amounts of heat; store and release greenhouse gases; undergo changes between sea-ice cover that is highly reflective of solar radiation and open water that efficiently absorbs solar irradiation; and because they provide all the water needed for the freshwater, or hydrologic, cycle of evaporation, atmospheric vapor transport, and condensation and precipitation.

Because of the large quantities of energy needed for evaporation and released upon condensation to form rain, the hydrologic cycle is a critical means for energy transfer through the atmosphere. This energy is known as latent heat, and it amounts to 2257 kilojoules, or 540 kilocalories, per liter, which is as much as the energy contained in three standard bars of creamy chocolate with caramel (named after a red planet) for every single liter of water that is evaporated or condensed. The strong uptake of latent heat during transformation of liquid water into vapor is familiar to us all by the way in which perspiration cools our bodies; when the sweat evaporates from our skin, heat is taken up, and this lowers our body temperature. Every year, an estimated 350,000 cubic kilometers of water evaporates from the world ocean, and this draws heat from the ocean surface to the equivalent of the energy contained in about one million trillion of such chocolate bars (that is, a one followed by 18 zeros). For another measure of scale, this is about 2000 times humanity's annual global energy consumption of recent years. And the same amount of heat is released elsewhere in the atmosphere during condensation to form precipitation within the same year; there can be no doubt that the hydrologic cycle moves an awful lot of energy. Hydrosphere influences in the climate system act over time scales that can be very short—for example, when evaporation reacts immediately to temperature change—to thousands of years, as is the case for warming or cooling of the entire ocean volume.

Over longer geologic time scales of several thousands of years, another water-related component comes into play. It concerns the growth and retreat of land ice, notably of great continental ice sheets, such as those seen today in Antarctica and Greenland. We collectively name these icy components the cryosphere. Ice sheets form over thousands to tens of thousands of years, and decay over thousands of years. Ice and, in particular, snow are more reflective to solar radiation than bare rock, grassland, or woodland. In consequence, growth of a major ice sheet means that more solar radiation onto Earth's surface is reflected straight back out to space without warming the surface, so that further cooling and ice expansion ensue. This is known as the ice-albedo effect or ice-albedo feedback.

On time scales of hundreds to many thousands of years, the biosphere (the life cover on Earth) also plays a direct role in climate change, since vegetation changes influence the reflectivity of Earth's surface to incoming solar radiation too. Lush woodland is less reflective than steppe or tundra, and those, in turn, are less reflective than a sparsely vegetated desert. These impacts are captured under the term vegetation-albedo effect.

This brings us to a point where a few less-familiar controls on climate change have to be introduced, for study of climate changes on long geologic time scales of hundreds of thousands to many millions of years. While the atmosphere, hydrosphere, and cryosphere are the traditionally considered components of the climate system, evaluations over very long time scales require that we view the climate system in a broader sense, which is often captured under the name "Earth system." It includes two additional components.

The first concerns carbon uptake and release related to processes associated with the biosphere. These carbon exchanges typically take place in the form of CO_2 or methane (CH_4), which are major greenhouse gases. The influences of biosphere processes on concentrations of these greenhouse gases are called carbon-cycle, or biogeochemical, influences.

The second additional component concerns interactions with the lithosphere, the domain of rocks and sediments. In essence, this is all about plate-tectonic interactions with the rock cycle of destruction (weathering) and formation of rocks. Weathering causes slow but relentless CO_2 uptake from the atmosphere, while CO_2 emissions into the atmosphere take place via faster volcanic outgassing (see figure 9). Through geologic time, changes in the balance between these uptakes and inputs with different time scales have driven fluctuations between (warm) high-CO_2 greenhouse climates and (cold) low-CO_2 icehouse climates over millions to hundreds of millions of years (lower panel of figure 11). We will encounter many examples of this. For clarity, note that the shorter variations on time scales of tens of thousands of years in the upper panel of figure 11 are superimposed on these

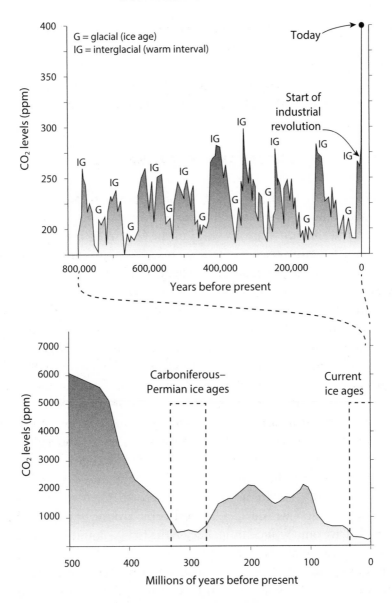

Figure 11. Schematic of long-term and shorter-term CO_2 variability through the past 500 million years, for which reconstructions exist. The last 800,000 years are known in detail owing to direct measurements of CO_2 in ancient-atmosphere bubbles trapped in ice cores.

longer changes, and relate more to carbon being shuffled around within the atmosphere-hydrosphere-biosphere system following pathways shown in figure 10.

After this familiarization with the main climate-system components, we're ready to evaluate the main drivers of climate change.

ORBITAL AND SOLAR CHANGES

In the late 1500s to early 1600s, Johannes Kepler laid the foundations of modern thought about the orbital relationships of planets revolving around the Sun. Isaac Newton in the late 1600s refined matters with his law of universal gravitation, and Albert Einstein again did so in the early 1900s with his work on general relativity. But, in general, Kepler's laws of planetary motion still provide good descriptions. Research into the influence of astronomical changes on climate had started in the 1800s, with particular progress by the French mathematician Joseph Adhemar and the Scottish self-taught physicist and astronomer James Croll. By the 1930s, the Serbian engineer Milutin Milanković (often represented as Milankovitch) had first calculated their impacts on the distribution of insolation, or exposure to solar radiation, over Earth's surface. The combined efforts of these pioneers, augmented by many others, laid the foundations of modern understanding of the astronomical drivers of climate change.

The key issues to be considered are changes in the shape of Earth's orbit around the Sun, in the angle of tilt of Earth's axis, and in Earth's seasonal position within its elliptical orbit around the Sun. These changes have only limited influence on the total amount of insolation that is received globally as averaged over a year, but to a greater extent control the seasonal changes and spatial distribution of that insolation over the planet. These so-called orbital changes in insolation occur in cycles that span several tens of thousands to several hundreds of thousands of years, and that are often referred to as the Milankovitch cycles. These cycles have fundamentally influenced Earth's climate and the oceans.

The great advantage of the orbital cycles and how they influence climate is that they have been rather stable through geologic time,

Figure 12. Photo of sedimentary variations imparted by the orbital cycles. (Photo by author; Punta Maiata, Sicily, Italy.)

and that they are measurably expressed in the fossil record of climate change. Often, they determine beautiful banding patterns in geologic sediment sequences, most commonly in sediments deposited in seas or lakes (figure 12). In the section I photographed for figure 12, the banding also plays a more mundane role, as a racetrack of sorts; youngsters on motor scooters race up atop one of the heavy white banks, and then try to brake as late as possible. The skid marks are impressive.

By recognizing the orbital cycles and knowing their time scales, we can use them as a tool for dating—in detail—sections of the geologic record where they are expressed, like using a metronome to keep time through a musical composition. These detailed datings can be verified against more sporadic datings from other methods—for example, based on dated geomagnetic reversals, or with radiometrically dated volcanic deposits, such as ash or lava, within the sediment sequence. The orbital time scales that can be developed and verified in this way are very useful for studying rates of change of processes in

Earth history. Orbital variability has been convincingly recognized in geologic sequences from very recent times to periods that date back to hundreds of millions and possibly more than a billion years ago.

As mentioned before, the yearly averaged influence of orbital cycles on insolation, and therefore climate, is small. But the seasonal changes in insolation and the distribution of insolation over the planet's surface have much larger impacts. Directly or indirectly, these drive responses in the Earth system such as ice buildup or melt, vegetation- and desert-zone changes, ocean-atmosphere exchanges of CO_2, and so on. These then act as feedbacks to the initial astronomically caused changes. A feedback process can be positive, in which case it amplifies the impacts of the initial change. It can also be negative, in which case the impacts are weakened. A large portion of the climate changes on time scales of tens of thousands to hundreds of thousands of years is initiated by the orbital changes in insolation, and then taken over by the feedback processes, sometimes to a point where the original orbital control is obscured because of the strong overruling influences of the feedback processes. But there are special mathematical techniques to filter out the underlying controls from fossil records of past climate changes, so that they can still be recognized.

To illustrate the way in which even relatively small astronomical changes can lead to big climate responses, consider that snowmelt in summer depends on a critical amount of heating from solar irradiation to create above-freezing temperatures. So in a critically sensitive setting, a small insolation change may make the difference between some snow surviving through the summer, or everything melting away. Most of us have seen an impact of small irradiation differences at work in the mountains: on the shaded side of mountains, snow survives much longer (often even through the year) than on the sun-bathed side of mountains at exactly the same altitude. With astronomical cycles, the difference that matters is not with respect to location, but in time. Insolation at our critically sensitive region goes up and down through time with the astronomical cycles. Typically, we are most concerned with 65° north latitude because this is close to the growth centers of the ice sheets that formed during ice ages. During periods with weaker insolation, some snow may survive the summer. In the next winter, it then grows bigger, so that more survives the subsequent summer, and so on. Thus, a permanent snowfield grows. This by itself then starts to affect the heat that is absorbed as well, because snow is highly reflective, and this feedback helps even more snow survive through the summers. After a long while, an ice cap forms.

The three fundamental cycles of astronomical change that influence climate are: the cycle of eccentricity of the Earth's orbit around the Sun, the cycle of precession of the equinoxes, and the cycle of obliquity or tilt of Earth's rotational axis relative to a perpendicular to the orbital plane.

The eccentricity cycle concerns changes in the shape of Earth's orbit around the Sun, due to variations in the gravitational pulls on Earth by the Sun, Jupiter, and Saturn. The orbit cycles between a near circular and a more elliptical shape every 100,000 years, and the amount of maximum ellipticity, or eccentricity, reached in each cycle increases and decreases over a period of 400,000 years. Though weak, the eccentricity of the orbit is sufficiently notable that a point is reached in the orbit where Earth is closest to the Sun (perihelion), and a point where Earth is furthest away from the Sun (aphelion). The direct impacts on the intensity of insolation that reaches Earth are weak, but more significant impacts result from interaction between the eccentricity cycle and the precession cycle. So let's first now look at precession, and thereafter at obliquity or tilt.

The cycle of precession of the equinoxes arises from a wobble of the Earth's rotational axis, like that of a spinning top, due to tidal forces between solid Earth and the Sun and Moon. On short time scales, Earth's rotational axis has a fixed direction in space, today pointing toward the star Polaris (the "North Star"). But on long time scales, this orientation changes because of the precession wobble. Half a precession cycle ago the axis was pointing toward the star Vega, a full precession cycle ago toward Polaris again, and so on. A full precession cycle takes 26,000 years, but the Earth's orbit also slowly rotates around the Sun, and the combined effects cause the precession cycle to express itself in insolation changes with two fundamental periods of about 19,000 and 23,000 years. The key impact is that the equinoxes and solstices slowly move around Earth's orbit. Today, the northern summer (southern winter) solstice almost coincides in the orbit with the position of aphelion, when Earth is furthest from the Sun. The northern winter (southern summer) solstice, in contrast, coincides closely with perihelion. In this configuration, northern seasonal contrasts are dampened, while southern seasonal contrasts are enhanced. Half a precession cycle ago, the opposite was true. A whole precession cycle ago, things were similar to today, and so on.

In times of minimum eccentricity, when Earth's orbit is almost circular, there is hardly any difference between the distances to the Sun at aphelion and perihelion; the distance would always be exactly the same in a perfectly circular orbit. In times of minimum eccentricity, the precession cycle therefore has hardly any influence. In contrast, times of strong eccentricity see a maximum difference between aphelion and perihelion, and the seasonal contrasts due to precession are amplified. The combined eccentricity and precession impacts are especially notable in seasonal contrasts, and in contrasts between the northern and southern hemispheres. We will revisit this when discussing monsoons, since those systems are particularly affected.

The obliquity cycle determines a cyclic variation in the tilt of Earth's rotational axis away from a perpendicular to Earth's orbital plane. Over 41,000 years, it changes from 24.5° to 22.1° and back again. Today, the tilt is just over 23.4°. The obliquity cycle has particular impact on insolation changes in the higher latitudes.

During the past 500,000 years, the maximum range of changes in total globally and annually averaged solar radiation at the top of the atmosphere amounted to only two watts per square meter, which is only a 0.7% variation in the total that reaches Earth. Taking into account a 30% reflectivity of the planet, as today, this implies a surface temperature variability over a range of about 1°C. However, spatial gradients of annually averaged, surface-absorbed solar radiation changed by up to seven watts per square meter, sufficient for changes in annually averaged temperature gradients of up to about 5°C. Such significant variations underlie a considerable part of Earth's natural climate variability. So, while orbital forcing was the ultimate driver of climate variability over tens of thousands to hundreds of thousands of years, it did not predominantly act through global annual mean variations. Instead, it acted by setting up spatial gradients of insolation. These, in turn, triggered variability in the dominant climate feedbacks, as outlined above, which resulted in the observed climate cycles.

Over very long time scales of hundreds of millions of years, the orbital periods of precession and obliquity have lengthened somewhat. We still don't know exactly by how much, but the periods have become longer, owing to a change in gravitational pull related to a slow increase in the distance of the Moon from Earth, along with a reduction in the rotational rate of Earth due to tidal friction. The latter has been quantified from geologic studies of tidal sediment deposits, which imply that Earth's faster rotation at around one billion years ago caused each day at that time to last only roughly 20 hours, relative to 24 hours today. Who would have thought it . . . in contrast to our mad-rush perception, the amount of time in each day is actually increasing!

While the orbital cycles are important for climate, they are not the only causes of fluctuations in incoming energy. Two further causes are related to variability in the Sun's output itself.

The first aspect of solar variability concerns the gradual increase in the Sun's luminosity, from only 70% of its modern intensity in the earliest days of Earth's history. This luminosity increase is extremely slow, with a gain of less than 7% of the modern intensity per billion years. So we should take the faint-young-Sun issue into account when considering changes in very deep Earth history, but we also need to recognize that solar output increase from this process is so slow that it is hardly relevant on time scales shorter than half a billion years.

The second aspect of solar variability concerns short-term fluctuations associated with sunspot cycles. Galileo Galilei is credited with the first systematic European sunspot observations in 1610, but the English astronomer Thomas Harriot also described them in the same year. Sunspots are darker, and therefore cooler (by 2000°!) regions on the Sun's surface. Sunspot activity is associated with less well-known activity of so-called "faculae," which are extra bright (by about 300°) areas on the Sun's surface. Faculae increase and decrease at the same time as sunspots. Although individual sunspots are larger in size and have a stronger temperature anomaly, there are many more faculae.

As a result, the effect of faculae on solar output just wins over the effect of sunspots, so that a sunspot maximum is marked by about 0.1% higher solar output than a sunspot minimum.

Sunspot activity varies from a minimum to a maximum and back again over about 11 years. In addition, there are superimposed cycles of variation in the intensity of the maxima; we call this amplitude modulation. Such an additional cycle, which is still well recorded in series of modern observations, has a period of 22 years. Other, less well-constrained periods of variability have also been reported—for example, at around 88, 210–230, and about 2300 years. Sometimes, a so-called grand minimum is reached, when sunspot activity drops to very low or near-zero values for an extended period of time. The most recent grand minimum took place between 1645 and 1715, and has been named the Maunder Minimum. There were two other grand minima close to that time—namely, the less intense Wolf Minimum of 1280 to 1350 and the strong Spörer Minimum of 1460 to 1550. Other groups of grand minima are known, and these groups seem to have occurred with a spacing of roughly 2300 to 2500 years.

The combined Spörer and Maunder minima coincided in time with a cool climate episode that has become known as the Little Ice Age. Of the three most recent grand minima, the Maunder Minimum was the most intense. In fact, it was one of the most intense grand minima of the last 12,000 years. Today, solar output is above average for that period of time.

There are several ways of reconstructing solar output for times before direct measurements, most notably using cosmogenic isotope ratios, such as carbon-14, beryllium-10, and—to a lesser extent—chlorine-36. These form high in the atmosphere by interaction of galactic cosmic rays with atmospheric gases. Solar wind deflects galactic cosmic rays, so a period with strong solar output is characterized by reduced cosmogenic isotope production. Carbon-14 production variations have been quantified in detail in support of the radiocarbon dating method, and beryllium-10 variations have been measured in ice cores drilled in Greenland and Antarctica. This work has revealed that even during the intense low of the Maunder Minimum, total solar irradiance was only 0.06%, to at most only 0.22%, lower than today. That implies reductions of globally and annually averaged irradiance of 0.15 to at most 0.5 watts per square meter at the top of the atmosphere. In terms of impacts on global mean temperature change, this implies a drop of 0.1°C to at most 0.35°C. These are small, but notable values.

Intriguingly, records of past climate and ocean changes show many cyclic variations with periods that are surprisingly close to those of the sunspot cycles—for example, in tree-ring records, laminated sediments, coral growth-band composition, and so on. This suggests that even small forcings of climate have somehow managed to trigger measurable responses, both in the atmosphere and in the oceans. But this is by no means fully accepted, as some climate-modeling studies have found that—even with invariant solar forcing—internal variability can appear within the climate system with time scales very similar to those of sunspot cycles, and that this may be mistakenly ascribed to sunspot variability. Whichever way this debate will be settled, one thing is clear: solar-output variations associated with sunspot cycles at best represent a weak component of climate forcing, relative to the dominant processes discussed throughout this chapter.

GREENHOUSE GASES

Greenhouse gases in the atmosphere partially block outgoing long-wave radiation from Earth to space. This outgoing long-wave radiation is thermal infrared radiation, the same as used in heat-vision equipment and thermal-imagery cameras. Earth, like any warm body, cools by radiating thermal infrared into its surroundings (space, in the case of Earth). The warmer the object is, the more intense the radiation, and thus the brighter the colors in thermal infrared images.

Warming of the Earth's surface due to incoming solar radiation causes it to emit thermal infrared radiation toward space, through the atmosphere. Changes in the concentrations of greenhouse gases then determine changes in the efficiency of Earth's heat loss to space. A useful analogy would be that these gases act like an insulating blanket around Earth, retaining heat and thus warming the surface. The climate system then starts to move toward a new equilibrium, as surface warming causes it to emit more intense thermal infrared radiation.

For a similar proportion of heat retention by greenhouse gases, this increased intensity then causes an increase in the loss to space. Over time, the net outcome is that, for a given increase in greenhouse-gas levels, Earth warms up to a point where the net energy loss again balances the net incoming energy; that is, until a new long-term energy balance is reached. If no balance were reached, then Earth would run away toward a permanent icehouse or greenhouse state. Given that Earth has spent virtually its entire history in the narrow temperature window in which liquid surface water could exist, we know that no such runaway has occurred, even if it has been a close call at times.

Warming due to solar-output variations will be noticeable everywhere in the atmosphere, both at the surface and at altitude. In contrast, the process of changes in atmospheric transmissivity to outgoing long-wave radiation by which greenhouse-gas increases affect climate causes a differentiated response, with warming concentrated near the Earth's surface alongside a tendency of cooling at high altitudes. This is one of several ways by which we can identify that present-day warming is a response to greenhouse-gas increases, rather than any sort of solar variability.

As an aside, today's heated debate about global climate change might create the impression that understanding of greenhouse-gas influences has been developed only recently. However, this is a serious misconception. As early as the 1860s, John Tyndall established experimentally that several gases in the atmosphere, including CO_2, absorb radiant heat. And in 1896, Svante Arrhenius specifically described the greenhouse effect of atmospheric CO_2.

Nowadays, we can measure the blocking of outgoing long-wave radiation using spectrometers on satellites far above Earth's atmosphere, a similar method to that used to measure the composition of atmospheres on other planets. The spectrometers measure radiation intensity over a wide range of wavelengths of radiation, and relate this to the temperature at which the radiation takes place. Dark bands of reduced output at certain wavelengths indicate enhanced spectral absorption by certain gases in the atmosphere, weakening the intensity of the escaping radiation. In other words, the transmissivity of the atmosphere is reduced at those wavelengths by the presence of certain gases.

In the thermal infrared bands at which Earth cools down, there is important spectral absorption at wavelengths typical for water vapor (H_2O), carbon dioxide (CO_2), methane (CH_4), and ozone (O_3), with smaller contributions at wavelengths for other gases. These gases hinder the escape of thermal infrared radiation and so cause retention of heat by the atmosphere; they are the dominant greenhouse gases. Without an atmosphere with these gases, Earth would be freezing at about $-18°C$, some $33°C$ colder than the global average that we observe. If you don't believe that a greenhouse effect exists or don't think that it is important, then try living and reproducing in a meat freezer; they operate at exactly the temperature that Earth would be without a greenhouse effect.

Water vapor contributes about half of the greenhouse effect. But water vapor is a peculiar greenhouse gas. It is a so-called condensing greenhouse gas, which will rapidly condense and fall out of the atmosphere as temperature drops, and evaporate to add vapor to the atmosphere as temperature rises. This makes water vapor a direct "follower" of any temperature changes (based on the so-called Clausius-Clapeyron relationship), and a powerful positive—amplifying—feedback, but it's not a primary driver.

CO_2, CH_4, and the other greenhouse gases are not condensing gases; their concentrations do not directly depend on temperature. Rather, temperature depends on their concentrations. Changes in their concentrations drive temperature changes, especially because these impacts are amplified by the water-vapor feedback. In addition, other feedbacks—such as ocean circulation and biogeochemical processes that are temperature sensitive—cause further increase or decrease in CO_2 and CH_4 levels within a complex interwoven system of variability.

It is useful to get a sense of the importance of greenhouse-gas changes for Earth's energy balance, and their potential implications for temperature, for comparison with those of orbital and sunspot changes. The impact on Earth's energy balance of the main noncondensing greenhouse gas, CO_2, in today's climate state scales to a global average increase of about four watts per square meter per doubling of the atmospheric CO_2 concentrations, or a similar decrease for every halving of the CO_2 levels. Over the ice-age cycles of the past million years, CO_2 levels changed between about 180 ppm during ice ages and 280 ppm during the intervening warm periods ("interglacials," like today before human emissions). Such a CO_2 increase scales to an extra $(100/180) \times 4 = 2.2$ watts per square meter of radiative climate forcing. By itself, that explains almost 2°C of the roughly 5°C global average warming from an ice age to the next warm period.

For deeper time, let's consider the Middle Cretaceous period, about 100 million years ago, when dinosaurs thrived. At that time, CO_2 levels were about 2000 ppm. We should consider this increase relative to an ice age in terms of doublings, so the first doubling is from 180 to 360, the second to 720, the third to 1440. The remaining $2000 - 1440 = 560$ ppm then represents a further $560/1440 = 0.4$ doubling. So, the Cretaceous value represents 3.4 doublings of CO_2 from the ice-age value. This roughly scales to $3.4 \times 4 = 13.6$ watts per square meter of extra radiative climate forcing relative to ice-age conditions. That alone might explain about 11°C of globally averaged warming, relative to an ice age. Several additional processes change these CO_2-only-based numbers, some increasing them and some decreasing them. So these numbers need to be considered with care; they represent only very simple estimates of the CO_2-based component of change. Also, the actual value of 2000 ppm for the Middle Cretaceous is under quite some debate, with more recent work tending toward lower values—but that is not so important here, as we were only doing a simple example calculation.

In the remainder of this section, we will evaluate some of the major natural controls on greenhouse-gas concentrations in the atmosphere-ocean system, which governed the natural climate and ocean variations noted in the geologic record. This system is well mixed for gases by a process of exchange between the ocean and atmosphere known as "equilibration."

Long-term controls on CO_2 concentrations are dominated by plate tectonics and the associated rock cycle. We evaluate these starting with the weathering of a simple example silicate mineral: $CaSiO_3$, or wollastonite. This weathering consumes CO_2 and produces dissolved positively charged ions, or "cations," of elements such as calcium, sodium, potassium, and aluminum, as well as bicarbonate ions (HCO_3^-)

and particulate or dissolved forms of quartz. These are then removed by groundwater and river water (see figure 9).

Weathering of a single unit (known as a "mole") of wollastonite consumes two moles of CO_2 and produces two moles of HCO_3^-. The groundwater and river water with those contents eventually flow out into the oceans, and our two moles of HCO_3^- are there used in shell formation by a wide variety of shell-bearing marine organisms. One mole of calcium carbonate is produced as part of a shell ($CaCO_3$), and one mole of CO_2 is released. By this stage, the weathering of one mole of silicate mineral has led to a net removal of one mole of CO_2, thus reducing the atmospheric concentration of that greenhouse gas.

Eventually the calcium carbonate shell gets buried in sediments on the seafloor, where it remains locked away from the atmosphere-ocean system. Millions of years later, the ocean crust with the calcium carbonate sediment is subducted. Under increased temperature and pressure in the upper mantle, the calcium carbonate sediment becomes chemically altered. So-called decarbonation reactions, similar to those in a lime-burning kiln, release our missing one mole of CO_2, which degasses into the atmosphere—likely via a volcano. The decarbonated remainder of our calcium carbonate (now calcium oxide, or CaO) readily reacts with other ions and notably silicon oxide (SiO_2, known as quartz in crystal form), producing one mole of silicate mineral ($CaSiO_3$). This mineral becomes part of the major tectonic evolution of subduction zones and gets squeezed and pushed up in the formation of mountain chains, such as the Rocky Mountains, Alps, Andes, or Himalayas. Eventually, the silicate minerals will become exposed to weathering, and that restarts the cycle.

In reality, there are many complications. Particularly important ones have to do with direct weathering of calcium carbonate before it has been transformed into silicates, and with burial and subsequent weathering of organic matter. But the fundamentals of the rock cycle outlined above dominate the main long-term fluctuations in CO_2 uptake and release over tens to hundreds of millions of years. This is because carbonate formation and its weathering, or organic matter formation and its weathering, more directly balance each other out for

CO_2. Only the formation and weathering of silicates show the time-separated processes of CO_2 balancing as described above.

Times of high plate-tectonic activity have increased rates of subduction, and the total length of the subduction zones may also be increased. This causes intensified degassing, which soon leads to rising CO_2 levels. In addition to the activity of plate tectonics, the intensity of the CO_2 input from degassing also depends on how carbonate rich the subducting sediments are. Today, the dominant Pacific subduction includes relatively little carbonate, and consequently the volcanic CO_2 emissions are relatively low. At around 100 million years ago, subduction was predominantly focused in a low-latitude ocean basin, called Tethys (see figure 4). Tethys contained large quantities of carbonate (reefs and suchlike), so that CO_2 emissions due to subduction were high. This added insult to injury with regard to CO_2 increases at that time, because plate-tectonic activity was also high.

Another consequence of enhanced plate-tectonic activity is an increase in mountain-range formation. This is a slow process that extends over many tens of millions of years because material is pushed and stacked both upward and, especially, downward, following which the less dense material that was forced downward into the more dense mantle will gradually rebound upward. This is the reason why young mountain ranges like the Alps, Andes, Himalayas, and Rockies are continuously being pushed up today, while material is being eroded and weathered off their surface. The consequently strong weathering acts to draw CO_2 from the atmosphere over tens of millions of years after the period of enhanced plate-tectonic activity.

Long-term CO_2 changes have also resulted from major changes in organic-matter storage in sediments, which are related to biological changes through time. At the same time, removal of large amounts of carbon from the atmosphere-ocean system—where it previously existed in the form of CO_2—released a great amount of oxygen into the atmosphere-ocean system. This helped the buildup from very low oxygenation levels to present-day levels and even higher, through Earth's history.

A good example of biological impacts on long-term CO_2 changes comes from the evolutionary changes that led to the Great Oxygenation Event of about 2.5 or 2.4 to 2.0 billion years ago. At that time, the rate of photosynthesis strongly increased, likely driven by cyano-bacteria. From that time onward, the atmosphere and shallow ocean waters seem to have contained small amounts of free oxygen, but the deeper oceans remained anoxic (no oxygen). A similarly remark-able period was the so-called Neoproterozoic Oxygenation Event of about 800 to 550 million years ago, which was associated with the emergence of many of the modern branches of photosynthetic eu-karyotic algae. During this event, free oxygen at last penetrated the deep oceans. More detail on this follows in chapter 4, since these pro-cesses were intimately related to global deep-freeze events known as snowball Earth.

Another great example of biological impacts on long-term CO_2 changes concerns the appearance of the first vascular land plants, roughly 420 million years ago. This type of plant possesses special-ized vessels for the transport of water and nutrients, and of the prod-ucts of photosynthesis. Its development represented a major step in the evolution of efficient land plants capable of colonizing Earth. In consequence, there followed a major development pulse of large forests and swamps over the next 100 million years or so. In swamps filled with water that had lost its oxygen to decomposition of plant matter, organic matter in the form of tree and plant remains became preserved and buried with sediment—if you want to see an example of this, then dig a bit in a swamp or peat bog. The buried remains from that period (the Devonian), and even more from the subsequent Carboniferous that lasted from about 360 to 300 million years ago, were transformed over time by being pressured and heated under the overlying sediment load. Thus formed the majority of today's com-mercial coal deposits.

Just before this period, CO_2 levels were very high—over 10 times higher than today (see figure 11). Removal of the staggering mass of Devonian-Carboniferous organic carbon from the atmosphere-ocean system meant that these high initial CO_2 levels dropped through the Devonian and Carboniferous, all the way down to values similar to those seen today. That is a roughly 10-times reduction of atmospheric CO_2 levels over about 100 million years, and it represents roughly four halvings of the CO_2 levels, from 5000 to 2500 ppm, to 1250 ppm, to 625 ppm, and then to a minimum of 300 or 400 ppm. Using modern relationships, that would represent a reduction in the energy balance of climate of some 16 watts per square meter, which equates to global average cooling of almost 13°C.

The removal of this vast amount of carbon from the atmosphere-ocean system released a great amount of oxygen. This caused a rise in atmospheric oxygen levels from about 15% before the CO_2 drop to about 35% at its end (for reference, today's levels are about 21%). For a long time, those high oxygen levels were thought to be the reason why the Carboniferous contains many fossils of giant insects. Insects have a primitive way of breathing via channels through their skin, which gets more efficient with higher oxygen levels. It was therefore suggested that increased oxygen concentrations allowed insects to reach bigger sizes, accounting for seagull-sized dragonflies, medium-dog-sized scorpions, and a 2.6-meter-long type of millipede. But more recent research attributes this to a lack of competition and predator pressure. Either way, this supersized insect fauna was not one that I would have enjoyed very much!

The major drop in CO_2 levels and its associated cooling helped usher in a major interval of ice ages during the Carboniferous and Permian periods, with the most pronounced phase between 320 and about 270 million years ago. So, while coal swamps were locking away massive amounts of carbon at low latitudes during the Carboniferous, the world gradually but steadily cooled down, and a major glaciation developed in the high southern latitudes. Ice sheets built up from growth centers on high mountains in the supercontinent Gondwana, which was centered on the south pole (see figures 4 and 5).

Finally, biological evolution has also affected the carbon cycle through evolving calcification strategies. This was referred to earlier when we introduced the Strangelove, Neritan, and Cretan ocean conditions. We will leave this aspect beyond consideration here, and instead pick up on it again when it's needed in the Reverberations section of chapter 4.

Volcanicity also plays a role in long-term CO_2 variations. Volcanoes emit CO_2. It's a notable amount, but actually not much in comparison to current human-induced CO_2 emissions—only a few percent. In the deep geologic past, there have been volcanic episodes of unimaginable scale and duration, which emitted vast quantities of CO_2 over typical time scales of one to several million years. Interestingly, their accumulated emissions may have been large, but the associated rates of CO_2 rise per year still remained low by comparison with current human-induced emissions. It was the persistence of enhanced natural rates over very long time scales that led to their large cumulative impacts.

We saw before that intense volcanicity occurs during periods of intensified plate tectonics. But the events considered here were exceptional even by those standards. These events concerned the outpouring of what is known as large igneous provinces (LIPs) or large volcanic provinces (LVPs), which are very large flood deposits of volcanic rocks known as basalt and andesite. Because it is more commonly familiar, I will stick with the word *volcanic* (hence LVP), but note that the professional literature favors *igneous* (hence LIP). LVPs formed at volumes and rates that greatly exceeded those of modern volcanic outpourings. Sites where crust rifts open to start formation of a new ocean basin commonly are located over major mantle hot spots, and LVPs are often—but not always—associated with such early rifting sites. A hot spot is a place where crust overlies a concentrated upwelling of hot mantle material, known as a mantle plume. Hot spots also exist away from plate edges; Hawaii is a hot-spot system, but a minor one that does not produce a flood deposit.

Among the largest known LVPs are the North Atlantic Igneous Province of 62–55 million years ago, the Deccan Traps of 66 million years ago, the Caribbean LVP that focused on about 90 million years ago, the Paraná-Etendeka Traps of about 132 million years ago, the Ontong Java Plateau of about 122 million years ago, the Central Atlantic Magmatic Province of about 200 million years ago, the Siberian Traps of about 250 million years ago, and the Mackenzie LIP of about 1270 million years ago (figure 13). Each of these represents an outpouring of one to several million cubic kilometers of basalt or andesite, which could cover an area the size of the United States in a layer several hundreds of meters thick.

Many of the major LVP events date to roughly the same times as large-scale extinction events or ocean anoxic events (OAEs). Those

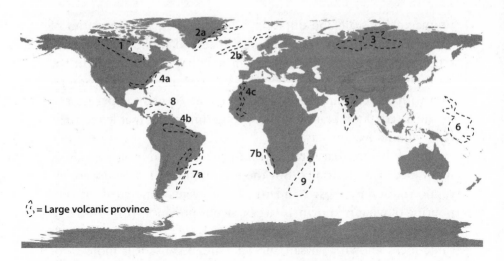

Figure 13. Map of the large volcanic provinces discussed in the text, as found in the world today. Note that the continental configuration has changed a great deal between the times these deposits were originally formed and the present, and also that more LVPs are known that do not feature in this book. 1, Mackenzie LIP (about 1270 Ma); 2a and b, North Atlantic Igneous Province (62–55 Ma); 3, Siberian Traps (250 Ma); 4a, b, and c, Central Atlantic Magmatic Province (201–199 Ma); 5, Deccan Traps (66 Ma); 6, Ontong Java Plateau (about 122 Ma); 7a and b, Paraná-Etendeka Traps (about 132 Ma); 8, Caribbean Flood Basalt (about 90 Ma); 9, Madagascar Flood Basalt (about 90 Ma). Ma stands for million years ago.

coincident events may therefore represent—in some way—responses to the massive volcanic emissions of CO_2 and sulfur dioxide (SO_2). We will revisit that relationship when we get to the OAEs.

Besides events of CO_2 release, there are strong indications that there have also been events of methane (CH_4) release, especially because of breakdown of so-called methane clathrates, which are also known as gas hydrates. These are complex structures in which CH_4 molecules are trapped within water-ice crystal structures that are stable at very specific temperature and pressure conditions.

Gas hydrates occur on land, in the permanently frozen soils called permafrost at high latitudes. They also occur inside seafloor sediments, at shallower depths (lower pressure) when the water is very cold, and deeper (higher pressure) when the water is warmer. Even though many people have never heard about them, gas hydrates are not rare; the total global amount of CH_4 contained in clathrates today is estimated to be 2 to 10 times the volume of all other known natural-gas sources combined.

Gas hydrates are always good for some researcher playtime. You can let the ice melt in your hand, and light up the CH_4 that is released. I have done this once or twice and find it great fun, but then I'm probably not always the most mature and serious person around (some colleagues get a bit cross with me, as they want to keep it all for analysis). Literally, it looks like a burning snowball—I just can't resist that.

It is beyond the scope of this book to further elaborate on clathrates, other than to say that ocean warming and permafrost melting are good ways of breaking up clathrates and releasing the CH_4 they contain. Also, increased heat flow below early spreading ocean crust can warm up sediments on top of that crust sufficiently to destabilize any clathrates they contain. And, finally, clathrates in submarine sediments may become destabilized when large submarine landslides occur, since removal of the slide's mass reduces pressure on the remaining (initially underlying) sediments. The breakdown of clathrates that follows can further destabilize the sediments, causing more movement. In going through the history of the oceans, we will encounter some instances of CH_4 release that were related to one or more of these processes.

PLATE TECTONICS

Besides controlling the rock cycle, plate tectonics also determines the spatial distribution of land and sea, and the locations and intensity of new mountain-range formation. Both are important for climate.

The location of continents on the globe affects the reflectivity of the globe. Open water is among the most absorbing surfaces for incoming radiation, especially at lower latitudes where the sun reaches very high angles overhead. Continental surfaces are much more reflective, especially when covered by desert, steppe, or grasses and shrubs, as they often are at subtropical lower latitudes. As a result, past times when most of the continental mass existed at lower latitudes would have been characterized by a stronger planetary albedo (reflectivity) than times when the continents were at high latitudes and oceans dominated the low latitudes. In addition, weathering rates are higher in warm and wet conditions than in cool and dry conditions, so that dominance of continents at low latitudes will result in more efficient CO_2 removal due to weathering. A configuration with continents clustered at low latitudes existed at the onset of the snowball Earth periods of 2.4 billion and 750 million years ago, and is thought to have helped push the climate system into deep freezes.

The locations and intensity of mountain-chain formation are important too, especially through the interaction between such mountain chains and the atmospheric jet streams. These jet streams meander around the globe, and their meanders can become fixed into big north–south swings by obstructions due to north–south mountain chains like the Rockies and Andes, or pushed permanently or seasonally southward or northward by major mountain chains at subtropical latitudes, like the Himalayas. Fixed swings in the jet stream lead to long-term constant climatic conditions, with one side of the swing often associated with strong drought. This (partly) explains long-term prevalence of certain major deserts, such as those in the southwestern United States, and the Gobi Desert and Chinese Loess plateau. Deserts are highly reflective to incoming solar radiation, and thus affect the energy balance of climate. In addition, the presence of high mountains offers suitable sites for initiation of ice-cap growth. These caps then expand downward and outward while the ice-albedo feedback increases cooling, until they coalesce into a continental ice sheet with a strong self-sustaining ice-albedo effect.

IMPACTS

Are asteroids and comets important for climate? The short answer here is: yes. There is no question that an impact by a large extraterrestrial body has devastating impacts. It is now well established that such an impact happened at the end of the Cretaceous period, when dinosaurs went extinct, 66 million years ago (see figure 1). This does not mean that nothing else was happening at around that time, independent from the impact. Indeed, at around the same time there was the Deccan Traps flood-basalt deposition, and some researchers make sound arguments that such additional changes may have contributed to the extinction event, by causing a longer-term underlying environmental deterioration. But then the impact occurred, and it is very well documented: for example, its crater has been found partly on land and partly offshore Yucatán; enrichments in iridium, an element typical for asteroids, have been found strewn around on an almost global scale from the impact site; and massive tsunami deposits have been documented around the impact site.

A comprehensive but continually debated extinction scenario has been proposed for this impact event, described as an impact winter, in analogy to the nuclear winter scenario that arose from studies of the potential aftermath of a global thermonuclear war. Other impact events at other times have been documented as well, but the end-Cretaceous one is the one that has been linked most closely, and beyond reasonable doubt, to the timing of a major extinction event.

CHAPTER 4

SNOWBALL EARTH AND
THE EXPLOSIONS OF LIFE

Several times in Earth's history, the planet has become completely covered in ice and sea ice, from pole to pole; we call this the "snowball-Earth" state. We start our discussion of climate and ocean events with snowball Earth and its intricate relationship with the evolution of life on the planet (see figure 1).

There have been several snowball-Earth episodes, the earliest taking place about 2.4 billion years ago. This means that, after the appearance of Earth's oceans at around 4 billion years ago, we have skipped in fast-forward mode through some 1.6 billion years of Earth history. We skipped this because there is remarkably little that we can say about continent-ocean configurations and ocean conditions through that immensely long interval, except that there was hardly any oxygen in the oceans, and that life was restricted to single-celled prokaryotes in aquatic environments (prokaryotes consist of small, simple cells without a nucleus or other organelles). Plate-tectonic reconstructions of the continent-ocean distribution through time cannot coherently look back in time beyond 750 to 800 million years ago. Although more information about life and ocean conditions through the early times keeps slowly but steadily emerging, notably from organic geochemical and isotopic studies, it is only from the first snowball-Earth episode that the geologic record provides information of a richer detail and diversity. In addition, the first major developments in life and oxygen availability occurred at around this time. For these reasons, I have selected snowball Earth as the starting point of our journey through the history of the oceans.

Still, our discussion of the snowball-Earth episodes lacks the depth and broadness that we will encounter with more recent events. This is a direct reflection of the fact that information remains scarce, even for these extreme events. The snowball periods are so old that only scattered geologic evidence exists from the most ancient cratonic interiors, and there is hardly any fossil record. Indeed, life was restricted to microscopic single-celled organisms and a few very primitive centimeter-scale multicellular organisms until right at the end of the youngest snowball-Earth episode. Then, life experienced two abrupt explosions of diversity and complexity. It is thought that these explosions were partly driven, or at least made possible, by the preceding snowball-Earth episodes. Some studies even suggest that the emergence of complex life may have been partly responsible for snowball Earth. As ever, deep interconnections are emerging between life and climate. This chapter first outlines snowball Earth, then the explosions of life, and finally some long-lasting implications of the presence of life in the wider Earth system.

INTO THE FREEZER

Snowball-Earth episodes represent the planet's closest shave with uninhabitable conditions, and might have culminated in totally barren conditions if Earth's plate tectonics and volcanism had been less vigorous. During snowball episodes, Earth almost ended up in a runaway, frozen, climate state. This is especially evident for two of the more recent events—the Sturtian of 720 to 660 million years ago and the Marinoan of 650 to 630 million years ago—for which sedimentary evidence of extensive glaciation has been found on every continent of the world bar Antarctica (the latter cannot be assessed because of its modern ice sheet).

Evidence is increasing that individual episodes lasted millions of years. They shouldn't be confused with so-called "icehouse states" of climate, which were longer periods with ice sheets in at least one

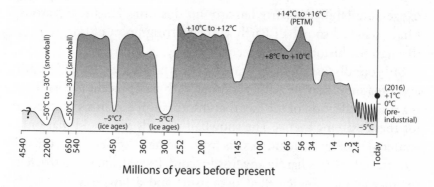

Figure 14. Temperature fluctuations through Earth's history. This is a highly schematic representation for indicative purposes only. Note the highly nonlinear time axis. All temperature values are presented as offsets from the global average temperature of preindustrial times. PETM stands for Paleocene-Eocene Thermal Maximum.

polar area, and which were marked by regular fluctuations between ice ages and warmer periods that are caused by Earth's orbital variations. By contrast, the snowball state is continually frozen over a very long period of time. For a completely (highly reflective) ice-covered world, energy balance calculations indicate that the global average temperature may have reached as low as −50°C (figure 14), with equatorial regions sitting at around −20°C.

Widespread ice cover and deep-freezing conditions would have terminated aquatic ecosystems on land (the only land-based ecosystems that may have existed at the time). Sea ice is thought to have been from a kilometer thick at high latitudes to a few meters in equatorial regions. A cap of floating sea ice of more than 20 meters thick would shut off light penetration, and thus photosynthesis, in which case marine productivity would collapse. The available evidence indicates a very strong global reduction in the marine ecosystem, but life clearly survived through the episodes overall. This—in conjunction with modeling experiments—is a key piece of evidence used to argue that sea-ice/icy-slush thickness in equatorial regions was (at least locally) less than 20 meters and more likely closer to 2 meters, allowing photosynthetic organisms to just survive the prolonged deep-freeze conditions. Some modeling studies suggest that there may even have been some low-latitude regions that remained largely ice-free, but this has not gained much traction with the mainstream community.

Snowball-Earth episodes were reconstructed from widespread findings of geologic deposits that reflect the actions of ice. These include scraping; bulldozing; freezing and thawing; transportation of rocks of all shapes, compositions, and sizes; and release of these as dropstones. The English geologist W. Brian Harland was the first to convincingly demonstrate, in 1964, that some of the glacial deposits originated from low latitudes, using measurements of magnetic directions recorded in the sediments. The American geoscientist Joe Kirschvink related theory to observations in the late 1980s and introduced the term snowball Earth.

Extensive sea-ice or icy-slush cover over the world ocean limited gas exchange between the atmosphere and oceans during snowball-Earth episodes. In consequence, remarkably large gas-concentration differences built up over time between air and sea. For example, we will see that snowball-Earth episodes started following intense reductions of atmospheric CO_2 levels, and air-sea gas exchanges in these times before sea-ice coverage would have ensured that CO_2 levels in the oceans were similarly reduced. As the planet subsequently froze over, atmospheric CO_2 started to slowly build up to extremely high values, owing to steady volcanic outgassing by volcanoes through the several-million-years' duration of snowball-Earth episodes. Atmospheric CO_2 levels toward the end of the snowball-Earth episodes are thought to have reached several hundred times the modern levels. Such dramatic buildup could occur because the planet's extensive ice cover curtailed weathering on land that would otherwise have consumed CO_2, and because the sea-ice cover prevented air-sea gas equilibration as well as biological production in the oceans, so that the vast potential of the ocean to absorb and store CO_2 remained unavailable.

Within the oceans, submarine volcanic CO_2 emissions were balanced by CO_2 consumption during submarine weathering of fresh volcanic rocks, which also drove intense calcium and magnesium enrichment. When the snowball's sea-ice barrier finally fragmented, biological production picked up again and massive carbonate deposition was triggered at the same time. The various processes together caused an overshoot to intense greenhouse conditions immediately after the snowball-Earth episodes.

The above sets the scene in broad brushstrokes. But we need to go into more detail to better appreciate the remarkable snowball-Earth events.

Two broad snowball periods have been inferred from the geologic record, although debate remains about whether the geologic record through each episode indicates just severe glaciations or true snowball conditions. Incidentally, such debate is normal and beneficial in the natural sciences because we can never obtain completely conclusive answers from the sparse available evidence used to characterize periods of time many millions to billions of years ago. It's the debate and associated quest for improvement that keep my colleagues and me going; we're like detectives trying to piece together an account of what happened from a web of incomplete fragments of evidence and indirect hints. Every new piece of evidence refines and/or slightly alters the reconstruction. Continuous communication with colleagues working on the same problem from a different perspective is essential, which is why paleoclimate researchers greatly value publications and discussions. Many of us feel conflicted by traveling and actively try to reduce our "carbon footprints," but the simple fact is that nothing beats frank face-to-face discussions at an international conference or workshop for advancing the field and training students.

The first snowball period took place between about 2.4 and 2.1 billion years ago, in the Paleoproterozoic era, and is known as the Huronian glaciation after the area where its major glacial deposits were first found (see figures 1 and 2). There may have been at least three separate snowball episodes within the Huronian.

The second snowball period dates to the Neoproterozoic era, and has thus become known as the Neoproterozoic snowball-Earth period (see figures 1 and 2). This snowball period is much younger than the Huronian, at 750 to 580 million years ago, so that much more geologic evidence has been preserved for us to study. Again, there were several separate snowball episodes within that broad period. First came the Kaigas episode at around 750 million years ago, but there is much debate whether this was a severe icehouse state or a true snowball state—the evidence is quite weak. So discussion of the Neoproterozoic snowball period focuses on the other two episodes, which are more widely accepted: the Sturtian episode of 720 to 660 million years ago, and the Marinoan episode of 650 to 635 million years ago. These two fall within a geologic period with the fantastically descriptive name of Cryogenian. Another, more minor episode (Gaskiers) followed at 582 to 580 million years ago, in the middle of the Ediacaran period that extended from 635 to 542 million years ago. During this Ediacaran period, complex multicellular life became well established, especially with a veritable "explosion" of body types shortly after the Gaskiers snowball-Earth episode. We will return to that after we have looked at the drivers behind the onsets and endings of snowball-Earth episodes.

Snowball-Earth episodes were most likely caused by a rare confluence of three factors: weak solar output, changes in Earth's reflectivity to incoming solar radiation, and greenhouse-gas changes.

We know that solar output was lower at the times of the snowball periods because of the faint-young-Sun issue that we encountered earlier. At the onset of the Huronian, solar output likely was roughly 15% lower than today. During the Neoproterozoic snowball-Earth episodes, it was likely about 5% lower than today. With respect to the energy balance at the top of the atmosphere, this translates to global annual average incoming short-wave radiation of about 290 watts per square meter for the Huronian, and 325 watts per square meter for the Neoproterozoic, respectively. These compare with 340 watts per square meter today. Such reductions set the scene for a much cooler Earth than the one we know. But it's what happened subsequently with the incoming radiation that made the real difference.

Changes in reflectivity have a key impact on solar radiation reaching the planet. Anyone who has been driving, skiing, or trekking through wintery landscapes will know that good sunglasses are required because snow and ice are highly reflective to incoming radiation. In quantitative terms, bare ice reflects about 50% of incoming radiation, while fresh snow reflects as much as 90%. So, between the maximum and minimum reflection of a snow-and-ice-covered world, only roughly 30 to 160 watts per square meter of incoming solar radiation would have been absorbed by the surface for the Neoproterozoic events (for interest, this is calculated as: $[1 - 0.90] \times 325$ to $[1 - 0.5] \times 325$). For the Huronian, with only 290 watts per square meter of incoming solar radiation, this sum works out to roughly 30 to 150 watts per square meter of absorbed radiation.

At such low amounts of absorbed radiation, the snowball-Earth state is stable. In other words, once in this state, Earth should stay permanently frozen. We can make a basic calculation to illustrate this, based on today's greenhouse-gas concentrations (with preindustrial CO_2 levels of 280 ppm). Energy-balance calculations show that 30 watts per square meter of absorbed radiation would in the modern (preindustrial) world lead to a globally averaged equilibrium temperature of only $-100°C$, and 160 watts per square meter to a temperature of roughly $-10°C$. A globally averaged temperature of $-10°C$ is considered to be prerequisite for tipping into the snowball state, since it is the level at which low latitudes are cold enough for ice to survive throughout the seasonal variability around the average.

The snowball-Earth concept was first elaborated using energy-balance calculations by the Russian climatologist Mikhail Budyko in 1969 and the American physicist Aron Faegre in 1972. They demonstrated that, in a time of extensive ice buildup toward the equator between 25 and 30 degrees of latitude, strong ice-albedo feedback would drive climate past a tipping point where total global glaciation becomes inevitable. This immediately refocuses the question about snowball Earth to one about how and why such an initial extensive glaciation could develop that drove the system to this tipping point. Key to the answer is the fact that, given the faint young Sun, Earth must have had intense greenhouse-gas-based retention of outgoing heat energy in times before the snowball event, or else it would have been a frozen ball already. Then, the high greenhouse-gas concentrations must have become sufficiently reduced to start the deep freeze that eventually crossed the tipping point into a snowball state. The combined findings of Budyko and Faegre illustrate that, once established, a snowball state may only be broken if the system is forced back through the tipping point. At the levels of reflection that characterize snowball Earth, this is simply impossible to do with incoming radiation. What is needed, therefore, is very strong retention of outgoing energy— again, this points squarely at greenhouse-gas changes.

So, understanding the causes behind the start and end of snowball-Earth episodes comes down to figuring out the causes of the greenhouse-gas changes. In this sense, the Huronian snowball-Earth period makes great sense, because it would have occurred around the time that notable free oxygen began to appear in the combined atmosphere and ocean system, the Great Oxygenation Event. Up to that time, very low oxygen levels—thought to have been less than one hundredth of present levels—would have allowed much more methane to exist in the atmosphere than today. Methane (CH_4) had accumulated over a long time, from CH_4-generating decomposition of organic matter. Even today, methanogenesis is critical in decomposition—it is the key reason why landfill sites must be vented to avoid buildup of so-called biogas (which is mainly CH_4), and why masses of CH_4 escape from wetlands, swamps, and rice fields. CH_4 is a very powerful greenhouse gas; comparing equal masses, it is some 30 times more potent than CO_2. In the absence of oxygen, CH_4 would not have oxidized to CO_2, as it quickly does today. As a result, there would have been high levels of CH_4 in the atmosphere before the oxygenation event that started in earnest at around 2.4 billion years ago. The oxygenation subsequently caused a sharp reduction in the methane concentrations through oxidation reactions.

There also seem to have been high levels of CO_2, which likely had built up over time because of volcanic degassing in the absence of efficient removal of CO_2 by photosynthesis. Photosynthesis splits CO_2 into carbon, used in formation of new plant matter, and oxygen. Consequently, the oxygenation event must signal a sharp reduction in CO_2 levels, in similar ways as discussed for the Carboniferous decrease of CO_2 and rise of oxygen.

It would be good to get a bit more specific estimate of the sort of greenhouse-gas concentrations that existed during the presnowball climate state, which was ice-free. We don't really need a precise number; instead, let's just do a crude calculation of what would have been needed to avoid ice buildup. We previously saw that absorbed incoming radiation in a snowball state amounted to between 30 and 160 watts per square meter, assuming a total global cover with fresh snow (likely on top of ice), or with darker ice, respectively. Let's assume a scenario roughly in the middle of the range, at 90 watts per

square meter. This roughly corresponds to a global average temperature of about −50°C, similar to results from more sophisticated calculations. To remain unfrozen, the times before snowball Earth must
at least have had globally averaged temperatures higher than −10°C.
To achieve that, at least 70 watts per square meter extra heat retention
would be needed.

Typically (under modern conditions, that is), half of the extra heat
retention would come from the water-vapor feedback effect that
strongly amplifies other changes. If we assume that this also applied
to snowball Earth, then this leaves about 35 watts per square meter to be explained by means of the other greenhouse gases. Now
let's make another rough (modern) assumption that half of this was
caused by more abundant but less potent CO_2, and the other half
by less abundant but more potent CH_4. Then our CO_2 contribution
would be about 18 watts per square meter, or four and a half doublings of CO_2 from the modern (preindustrial) values of 280 ppm
that we started our argument with. Our crude calculation therefore
suggests presnowball CO_2 levels of 7000 ppm or more. This rough
estimate compares well with some published values, although much
higher values have also been suggested.

In short, we have now seen that greenhouse-gas concentrations
outside snowball-Earth episodes were very high. We have also discussed how reductions in those levels would have caused a deep freeze,
which eventually crossed the tipping point into a snowball state owing to the strong ice-albedo feedback. Finally, we have seen how the
oxygenation event that started at around 2.4 billion years ago provides a sound indication that the Huronian snowball episode indeed
followed dramatic CH_4 and CO_2 removal. This raises the question
whether a similar sequence of events can also be convincingly demonstrated for the Neoproterozoic snowball-Earth episodes.

Indeed, there are good indications for CO_2 reductions leading into
the Neoproterozoic snowball-Earth episodes. One indicator is a gradual increase in the proportion of carbon-13 relative to carbon-12; this
is quite a common signal before most snowball episodes. These are
two stable isotopes of carbon, where carbon-13 has one more neu-
tron in its atomic core than carbon-12. Given that reactions of carbon removal commonly favor carbon-12 over carbon-13, the gradual

carbon-13 increase signals CO_2 removal from the atmosphere-ocean system.

If part of the CO_2 removal was caused by photosynthesis, then improved oxygenation and consequent further CH_4 removal may also have occurred. But the main CO_2 reduction has been ascribed to phases of intensified weathering, of fresh volcanic rocks among other things, between one billion and 540 million years ago. The period marked a multiphased breakup of a supercontinent known as Rodinia, as new ocean basins developed. Development of extensive lengths of new spreading ridges in initially shallow water promoted an explosive kind of volcanism that creates great volumes of fragmented fresh volcanic rock. This delivered masses of rock that weathers easily, in a form with many edges and corners for chemical attack, which is thought to have been an important source of calcium and magnesium release into the oceans, along with marine CO_2 consumption. One of the newly opening ocean basins, which would eventually become the Pacific between the North American and the Chinese and Australian-Antarctic blocks, went through a first phase of opening at 750 to 630 million years ago. Another basin (called the Iapetus Ocean) began to open at 615 to 580 million years ago, between the North American and the central Asian and European blocks (see figure 4). If you're keen, the studies by Z. X. Li et al. or Gernon and colleagues in the Key Sources for Chapter 4 in the Bibliography offer a good entry point for plate-tectonic summaries of these times. As I said before, I refrain from including my own sketches because there remains much uncertainty.

Weathering was also promoted by the uplift of a great mountain chain, of similar scale to the Himalayas, from around 600 million years ago. Finally, CO_2 extraction through weathering was efficient at the time because the Rodinia supercontinent was located at low latitudes, where warm and humid conditions cause high rates of weathering, while the opening of nascent ocean basins in between separating continents triggered intensification of evaporation-precipitation cycles.

In another fascinating tale of how life may shape climate, recent work has suggested that biological evolution, and in particular the rise to prominence of eukaryotic algae, may have inadvertently contributed to triggering of the Neoproterozoic snowball-Earth episodes. This work considers that, besides relatively weak solar radiation and changes in greenhouse gases, changes in our planet's reflectivity due to cloud cover also strongly affect the radiation balance, and the rise of eukaryotic algae influenced cloud cover.

Cloud vapor forms through condensation, and condensation is promoted by the presence of tiny impurities in the air that act as condensation nuclei. Modern examples include cloud seeding, which artificially triggers cloud nucleation by spreading an aerosol from airplanes into humid air, and formation of vapor trails—or contrails—on tiny particulates emitted from airplane engines. In nature, dimethyl sulfide, or DMS, is a key cloud-nucleation agent. It results from products emitted upon deterioration and decay of eukaryotic algae. Dilute DMS in air gives rise to the "healthy sea-air" smell, but in high concentrations it smells like rotting cabbage and truffles, and it can make a seaside stroll past abundant rotting seaweed seem rather less invigorating. Today, the global sulfur-aerosol injection into the atmosphere from DMS and sea spray combined is larger than that from volcanoes, so it's an important process.

DMS production would have become important when eukaryotic algae rose to ecological prominence at about 800 to 750 million years ago, after having lurked about in low abundances since 1.5 billion years ago or even earlier. With the associated rise of DMS production, cloud reflectivity would have become much more prominent. Climate-model experiments with low CO_2 indicate that this could drive a very rapid transition into snowball conditions. So all the prior discussion about intensive CO_2 reduction remains essential, but the superimposed DMS mechanisms may explain why a snowball state was rapidly triggered under such conditions. It's one of those ultimate ironies, in which life went through an evolutionary step change, only to cause conditions that might then have wiped it out. But, fortunately, it didn't quite complete that trajectory. Instead, the planet managed to escape from its frosty straitjacket. In doing so, it shaped the perfect conditions for an explosion of life. I am endlessly fascinated by such interactions; they exemplify the complex nature of the Earth system, full of intricate connections and feedback, and illustrate how disruptions to seemingly irrelevant components may eventually have major consequences.

OUT OF THE FREEZER, INTO A GREENHOUSE

We have seen that greenhouse-gas decreases were among the most important causes of the various snowball-Earth episodes. Greenhouse gases, again, are the main suspects for breaking Earth out of its frozen stranglehold.

To terminate a highly reflective snowball state, colossal atmospheric CO_2 concentrations are needed. These need to provide the additional heat retention to drive globally averaged temperatures up far enough to push the system back through a tipping point beyond which a normal climate state is stable. Climate-model studies estimate that CO_2 concentrations had to rise to well over 400 times the modern preindustrial value for this, to 130,000 ppm.

The suggested accumulation of extreme atmospheric CO_2 levels requires that normal CO_2-removal processes of absorption into the ocean, marine and land-based biological production, and weathering must have been largely switched off for a very long period of time (the duration of the snowball episode). Thus, CO_2 could build up because of ongoing volcanic emissions over the course of millions of years. The feasibility of this concept can be checked using carbon isotope ratios. Following the lead-up periods with increasing proportions of carbon-13 that we discussed earlier, snowball episodes themselves were associated with remarkable shifts to increasing carbon-12 proportions. Although debate remains (as ever . . .), it is thought that these shifts largely reflect a near-total collapse of biological production upon the onset of the deep freeze; biological removal of CO_2 preferentially involves carbon-12, so that inhibiting it would leave carbon-12 in the system. The shifts were strengthened by inhibition of CO_2 removal through weathering, due to the global ice cover and worldwide cold conditions of snowball Earth. Meanwhile, the sea-ice cover limited CO_2 absorption and storage in the oceans.

Once broken out of the snowball state, Earth still had an atmosphere that was absolutely loaded with CO_2. As a result, the snowball state straight away changed into a supergreenhouse state, with globally averaged temperatures up to +30°C or +40°C, compared with today's value of +15°C. This hot, wet world with extreme CO_2 concentrations, and with lots of pulverized rock because of the grinding and freezing actions of the preceding global glaciation, was characterized by very high rates of chemical weathering and associated CO_2 consumption. This weathering pulse led to intense ion transport into the oceans, enhancing the strong accumulations of calcium and magnesium ions in the oceans that had occurred throughout the duration of the snowball episodes owing to submarine weathering of fresh volcanic rock at new, shallow spreading ridges. This cocktail helped trigger the formation of anomalous carbonate formations during the post-snowball periods, known as the cap carbonates, which in places reach thicknesses of hundreds of meters. In spite of these strong carbon-removal processes, it still took millions of years to reduce the high CO_2 levels and settle the climate back from a supergreenhouse to a more normal state.

Strong input of weathering products also provided the oceans with much phosphorous, as witnessed by high phosphorous abundance in sedimentary iron formations of the time. Phosphorous is a critical nutrient in the oceans, and its buildup to high levels, along with an abundance of CO_2, fueled a strong increase in photosynthetic algal and bacterial production immediately after the snowball periods, and thus a boost in oxygen production.

Such a marked increase in oxygenation happened after the Huronian snowball-Earth period, but it did not persist. Instead, the initial rapid oxygen increase at that time seems to have withered away. Why did the trend not continue? Some researchers blame extinction events at around the time of two very large asteroid impacts for "resetting" the system to a low-oxygen (and relatively high-CO_2) mode. One was the massive Vredefort impact of 2.023 billion years ago in present-day South Africa, and the other was the equally enormous Sudbury impact of 1.849 billion years ago in present-day Canada.

After the Neoproterozoic, a rapid oxygen increase again occurred, toward levels of about 15%, and this time there were no major catastrophes. In consequence, oxygen now started—for the first time—to penetrate throughout the deep ocean. The culmination of this process occurred at around 551 million years ago, based on a massive carbon isotope shift that is thought to reflect oxidation of masses of dissolved carbon in the deep seas. This oxygenation set the scene for an explosion of life throughout the ocean.

A TALE OF TWO EXPLOSIONS

Toward the end of the Neoproterozoic snowball-Earth interval, in the Ediacaran Period that spanned the interval of 635 to 542 million years ago, something amazing occurred in the oceans. After billions of years of only simple single-celled and primitive multicellular life, much larger and more complex multicellular organisms burst onto the scene: the Ediacaran biota, named after the Ediacaran hills in South Australia. We speak of biota because we don't know what the organisms exactly represent. They may have been plants, animals, fungi, microbial communities, or intermediates between plants and animals. In the latter half of the Ediacaran Period, some of the biota reached meters in length.

Since the Ediacaran biota were entirely soft bodied, they did not fossilize very well. This means that we are lucky to have the fossils at all, and that they are hard to study. The Scottish geologist Alexander Murray is credited with the first discovery of fossils of Ediacaran biota, in 1868. In the late 1960s, better-preserved fossils were discovered in Newfoundland, Canada. The presence of microbial mats often helped preserve the soft-bodied organisms in settings where none of the biota would be preserved today.

The oldest microscopic fossil of Ediacaran biota dates to within a few million years of the end of the Marinoan snowball-Earth episode at around 630 million years ago, although there are claims of findings at levels even dating back to 770 million years ago. The oldest macroscopic fossil of Ediacaran biota dates back to about 610 million years ago, and the first varied community dates back to about 600 million years ago. It is clear, therefore, that these organisms had appeared well before, or relatively soon after, the Marinoan snowball-Earth episode, and definitely before the extensive Gaskiers glaciation of 582 to 580 million years ago.

After the Gaskiers glaciation, however, the Ediacaran fossils reveal a sudden major increase in diversity of body types at around 575 million years ago, followed by rapid increases in the complexity of each body type. This was the Avalon explosion, named after the Avalon Peninsula in Newfoundland (see figure 1). Similar strong Ediacaran diversification has been found in other places around the world, and in particular in northern Russia (White Sea assemblage) and Namibia (Nama assemblage). This supports the notion that this event was an explosion of life, similar to the later Cambrian explosion. In all, organisms appeared that looked like modern jellyfish, sea pens, worms, soft corals, sea anemones, and soft-bodied versions of arthropods.

Incidentally, many people—myself included—find that ancient body types are easier to grasp when expressed in analogies to modern shapes and life-forms. But caution is needed; we don't know (at all) if the resemblances we perceive signal true ancestry. Most researchers now think that the Ediacaran biota represent a partial or complete dead end, in which case the analogies are nothing but a reflection of our brains' tendency to associate unknown things with things we know already.

The Ediacaran biota largely seem to have gone extinct at around the time of a second rapid increase in biodiversity, at around 542 million years ago. Some of the Ediacaran biota may have left descendants. Regardless, at that fateful moment about 542 million years ago, things changed in a big way. Marine life underwent another explosion of diversity, this time with lasting success. Known as the Cambrian explosion, this event continued over 20 to 25 million years (see figure 1). Most modern branches (or phyla) within the kingdom Animalia abruptly emerged during this period. However, some early representatives of a few lineages had already been around before the Cambrian explosion, such as tube-forming worms, jellyfish-like organisms, and spongelike organisms dating back to 580 million years ago.

As early as 1841, the theologian, dean of Westminster, geologist, and paleontologist William Buckland—the same who first, in 1824, described a fossil animal that would later be named dinosaur—was among the very first to recognize the abrupt and transformative nature of the Cambrian explosion. But its great extent was particularly borne out by the excellently preserved fauna in the Burgess Shale (in the Canadian Rockies), which were discovered by the American paleontologist Charles Walcott. He collected tens of thousands of specimens between 1909 and 1924.

The Burgess Shale dates from about 508 million years ago and has exceptionally preserved even soft body parts. This rare condition attracted many more researchers to work on the Walcott collection, the Burgess Shale, and deposits with similarly excellent fossil preservation worldwide. The Burgess Shale is not the only known deposit with exceptional preservation of fauna from the Cambrian explosion. Other key examples are the Maotianshan Shales from Chengjiang, China, which are about 520 million years old, and the Sirius Passet deposits from Johan Peter Koch Fjord in far northern Greenland, which are about 518 million years old. Further major contributions have come from research on limestone nodules called Orsten, with especially important Cambrian examples found in the United Kingdom, Australia, China, Russia, Sweden, and Poland.

Notable among the pioneering researchers who worked on the Burgess Shale were the paleontologists Harry Blackmore Whittington, Simon Conway Morris, and Derek Briggs, who formally described and named many of the biota through the 1970s to 1990s. One of these was the enigmatic *Hallucigenia*, which was originally reconstructed upside down and back to front because of its odd shape (an elongated tube with tentacles/legs along one side and spines along the other), until new findings of greater detail revealed its real orientation.

Less than 15% of the Burgess Shale fauna contains hard body parts, which under normal conditions are the only parts that fossilize. Soft-body-part preservation was essential for appreciating the full diversity of animals that had arisen in the Cambrian explosion. The Burgess Shale fauna includes some swimming creatures, as well as mostly free-roaming or fixed bottom-dwelling creatures. Only about 10% of the species appear to be predators or scavengers, but predators and prey seem more equally distributed in terms of estimated biomass.

The appearances of complex multicellular life-forms in the Ediacaran and later the Cambrian represent vital step changes in the evolution of Earth's complex marine and land-based ecosystems, and thus also in the cycling of nutrients, carbon, and oxygen through the Earth system. One of the most obvious impacts was a major oxygenation event, which—for the first time since the oceans were formed—brought free oxygen into the deep sea.

There is no doubt that larger, mobile animals require the availability of oxygen for breaking down the ingested food, to sustain their more energy-expensive way of life. In that sense, it is not hard to see a likely relationship between the emergence of sufficient free oxygen and mobile animals. But this does not necessarily mean that oxygenation must have driven the appearance of multicellular animals. It cannot yet be excluded that the eruption of complex multicellular life-forms resulted from nothing but random chance in the evolutionary process. Also, there is no guarantee that there were no earlier increases in complexity that simply have not been preserved in the fossil record, or have not been discovered yet. If earlier increases in complexity did take place, then the "explosion" of life would have been more gradual than we now think. In that case, the currently inferred coincidence with oxygenation would be fortuitous.

In another twist of the tale, it has recently been suggested that the first animals, sponges, may actually have been instrumental in oxygenating the ocean. Sponges are among the most primitive animals, which evolved around 850 to 635 million years ago. They lack tissue differentiation—they don't have different types of tissues and organs. Modern sponges can live in waters with extremely low oxygen concentrations. If ancient sponges had the same capacity, this may explain their population of poorly oxygenated upper water layers during Neoproterozoic times before deep-sea oxygenation. Sponges are filter feeders that feed by filtering particulate matter from the water column. This filtering promotes water clarity, and the improved clarity allows more intense and deeper light penetration. This in turn promotes stronger photosynthesis by cyanobacteria and—notably—the eukaryotic algae that also developed into prominence at around these times. Sponges likely played a direct role in this change of photosynthetic communities too; their filter-feeding behavior preferentially removed small phytoplankton like cyanobacteria, and thus selectively increased the population of larger eukaryotic algae.

Because sponges consumed masses of organic particles in the upper water layers, less organic matter transferred into deep waters, where its decomposition would otherwise have depleted oxygen and released nutrients. The resultant drop in oxygen demand in the deep sea helped promote deep-sea oxygenation. The reduced release of nutrients into the oceans helped to curtail primary production in general. This, in turn, drove a general, long-lasting drop in oxygen demand for decomposition of sinking organic matter. Combined, these changes may explain much of the buildup of oxygen in the deep sea that we now recognize as a great oxygenation, and thus to the creation of oceanic environments that were suitable for occupation by mobile animals.

To me, this seems a compelling scenario, which will take some testing and validation. If it holds, then it is a magnificent example of interwoven complexity within the system. It has many parallels to "natural experiments" today, in which an initial environmental perturbation (such as introduction of a foreign species) triggers reverberations and feedback through the entire physical-chemical-biological system, very much like the suggested sponge-mediated adjustments. Next time you take a bath, give your sponge some thought. Its unassuming ancestors may well have played a critical role in shaping the planet such that you and I can now exist.

Two further implications of the changes at the time of the Neoproterozoic Oxygenation Event need to be briefly mentioned. First, the advent of larger, mobile grazing animals spelled the end of a billion-year (or longer) dominance of microbial mats. Today, stromatolites are relegated to a marginal existence in just a few specific locations. Together with the general decline of planktonic cyanobacteria at the Neoproterozoic Oxygenation Event, this drove a fundamental reorganization of the base of the oceanic food web to one dominated by eukaryote photosynthesis.

Second, the advent of intensified predator-prey interactions between mobile organisms is thought to have set the scene for the highly efficient "coevolution" mechanism. Coevolution occurs when predator pressure drives prey to develop defenses, which predators then develop countermeasures to, which prey then develop new defenses against, and so on. This leads to accelerated rates of evolution in both prey and predator organisms, and may underlie at least part of the rapid diversification that characterizes the two explosions, most notably the Cambrian.

REVERBERATIONS

Starting with the Cambrian explosion from 542 million years ago, the fundamental components of modern marine food webs were set up, and much of the oceans had become oxygenated. All the truly transformative ocean changes—from the alien early oceans to increasingly more recognizable oceans—had been completed. Some 3.5 billion years after they first appeared, therefore, the oceans and life within them had at last reached a stage where systems started to take shape that are still dominant today.

From the Cambrian onward, changes were subtler than before, and generally focused on shifts in the cycles of carbon, oxygen, and nutrients (phosphorous, nitrogen), including an especially notable change in the carbon system toward a truly modern type of operation from about 252 million years ago (and especially since 170 million years ago). In the following, we're going to need to spend some time discussing the basic controls on these cycles, as they are critical to all aspects of ocean and climate change, including the life that is sustained.

The key to it all is an interaction between biology (the organic carbon cycle) and geology (the inorganic, or mineral, carbon cycle). This complexity can cause some headaches when studying the ways of nature. But—in its defense—it is also exactly what makes nature so fascinating, and so resilient and capable of compensating for disruptions in separate facets of its interlocking processes. Critically, the interlocked system with its internal checks and balances is precisely why conditions on Earth have always remained suitable for organic life.

Although organic tissue is often summarized as CH_2O, we need to look at a more precise representation of the average composition of algal matter to understand the importance of nutrients. This is closer to $(CH_2O)_{106}(NH_3)_{16}H_3PO_4$ or, in summary form, $C_{106}H_{263}N_{16}O_{110}P$. The key message from these formulae is quite simple: for every 106 atoms of carbon (C), algal matter on average contains 16 atoms of nitrogen (N), and 1 atom of phosphorous (P). These main nutrients, or macronutrients, are essential for the formation of organic matter by photosynthesis.

There are also essential micronutrients, which are nutrients that are needed only in minute trace amounts. The most important one is iron (Fe). If any of the major or trace nutrients is in insufficient supply, then an organism cannot build organic tissue through photosynthesis—we call this a limiting nutrient. You may have heard in the news about iron-fertilization experiments. In these, researchers go to places where iron is the limiting nutrient, and scatter iron into surface waters to trigger production. Nature also has its own iron-fertilization mechanisms; notably, input of windblown continental dust that commonly contains abundant iron.

A ratio of C:N:P = 106:16:1 is, on average, observed throughout most of the ocean, especially in deep waters below about 500 meters. It has become known as the "Redfield ratio," after the American oceanographer Alfred Redfield who first described it in 1934. It is so characteristic in deep waters because of the decomposition of dead planktonic matter in those waters, which releases the components into the water column in the ratio that they were incorporated into the plankton. Ocean biochemists often measure whether a region's C:N:P ratio is roughly the same as the average Redfield ratio. Any significant differences are then used to guide investigations into regional deviations from regular biochemical processes.

To see how the organic and inorganic cycles affect each other, we're going to start with the organic side, using the simplified representation of organic matter: CH_2O. We saw previously that primary production through photosynthesis follows the pathway $CO_2 + H_2O \rightarrow CH_2O + O_2$, and the discussion above illustrates how nutrients are also included. Decomposition of organic matter, or respiration, follows the opposite direction: $CH_2O + O_2 \rightarrow CO_2 + H_2O$, which also releases nutrients. In the oceans, photosynthesis happens in shallow waters, where sunlight penetrates. So CO_2 and nutrients are extracted from surface waters, and O_2 is produced. Surface waters are in active gas exchange with the atmosphere, so that atmospheric gas levels change in response to surface-water gas levels (and vice versa). In terms of nutrients, nitrogen is seldom limiting, as there is a lot of it around; air is more than two-thirds nitrogen and exchanges with surface water, and some plankton can access dissolved nitrogen in water when nitrate is in short supply. For phosphorous and iron, however, external inputs are needed; for example, from rivers and/or windblown dust. If any nutrient becomes limiting, then this will curtail further production.

After death, decomposition takes place. In the oceans, most of this recycling of organic matter swiftly happens in the shallow waters, but some of the organics drop out of the photic layer into deeper waters—this is known as the biological pump. Deep waters are not in direct gas exchange with the atmosphere, so that decomposition in deep waters causes a steady decrease in oxygen and a steady increase in CO_2 and nutrients. Also, the deep waters become enriched in nutrients because of the general lack of primary production in the dark deep sea.

Eventually, ocean circulation brings deep waters back toward the surface in upwelling regions, where gas exchange with the atmosphere resets the deep-water properties into equilibrium with the atmosphere: effectively, CO_2 is released into the atmosphere, and oxygen is taken up from the atmosphere. At the same time, the upwelling deep water's nutrient load will trigger a feast of production; this is a well-known feature of upwelling regions. But note that, for oceanic deep waters today, reconnection with the surface through upwelling can take many hundreds to over a thousand years, and sometimes in the past it has taken even longer. The longer a deep-water mass remains isolated from gas exchange (in other words, the "older" the deep water gets), the lower its oxygen levels and higher its CO_2 and nutrient levels become, owing to ongoing decomposition.

We saw before that oxygen can only get into the deep sea through downward circulation of water from the surface, where it had exchanged gases with the atmosphere. In other words, deep-sea oxygen is replenished only through new deep-water formation. There are no other means of oxygen supply into the deep sea.

CO_2 that accumulates in the deep waters awaits one of two fates. One option is the aforementioned upwelling of the water mass to the surface, where it degasses during air-sea gas exchange. The other option involves a much slower, but on longer time scales very important, process—namely, interaction with seafloor sediments.

To look at this, we need to start from the way things work in the modern, Cretan-style ocean, which includes large-scale open-ocean carbonate formation by microscopic phytoplankton (algae) and zooplankton (unicellular "animals") that live free-floating in the water column—the planktonic calcifiers. These planktonic calcifiers only started to become established after the end-Permian extinction event of 252.3 million years ago (evaluated in chapter 5), and had risen to prominence by about 170 million years ago. Discussion of the modern functioning of the Cretan ocean sets an essential baseline from which we can consider the implications of an absence of planktonic calcifiers during the Neritan-ocean period between the Cambrian explosion and the end-Permian extinction event, and in the even older carbonate Strangelove oceans that dominated before the Cambrian explosion.

In today's (Cretan) oceans, marine sediments that are deposited at a depth shallower than about 4.2 to 5 kilometers commonly contain a lot of carbonate, or $CaCO_3$, because of accumulation over time of the hard skeletal parts of planktonic calcifiers. The chalk of the stunning white cliffs of Dover is a familiar example of such carbonate-rich marine sediments. Carbonate-forming algae live in surface waters, in the photic layer. Most carbonate-producing zooplankton also live in the productive photic layer, or in the first few hundred meters below it. From these productive regions, both organic matter and carbonate skeletal parts rain down into the deep sea. A lot of the dead organic matter is consumed and decomposed before it gets to the seafloor, releasing CO_2 into deep waters. Carbonate more effectively sinks to the seafloor, although a fair amount of it is dissolved on the way down as well.

This brings us to a crucial junction at which the organic carbon cycle interacts with the inorganic (mineral) carbon cycle. The formation of carbonate uses calcium (Ca^{2+}) and bicarbonate (HCO_3^-) ions. In simplified form, this is represented by: $Ca^{2+} + 2HCO_3^- \rightarrow CaCO_3 + CO_2 + H_2O$. Simply reverse this for carbonate dissolution: $CaCO_3 + CO_2 + H_2O \rightarrow Ca^{2+} + 2HCO_3^-$, which shows that an increase in CO_2 causes an increase in carbonate dissolution. Technically, this is because CO_2 in water forms the weak carbonic acid, which dissociates into HCO_3^- and H^+. This implies an increase in the water's acidity and thus in its corrosiveness to carbonate. So, as the deep-water CO_2 level goes up—notably, because of decomposition of organic matter—the water becomes corrosive and starts to dissolve carbonate. This affects both carbonate sinking through the water column, and carbonate that has reached the seafloor.

The intensification of carbonate dissolution with increasing deep-water CO_2 levels is key to a process called carbonate compensation. This term refers to the interaction—a negative feedback—between the oceanic carbon cycle and the underlying sediments, which tends to stabilize atmospheric CO_2 fluctuations on time scales of thousands of years. We can imagine its operation based on two key factors. First, sediments from the deepest parts of the oceans have no carbonate, while sediments from shallower waters do. Second, the depth at which dissolution of carbonate exactly balances the rain of carbonate from above, so that none accumulates in the sediment, is known as the carbonate compensation depth, or CCD. It was first described by the Scottish oceanographer John Murray in 1912.

When the ocean functions in a long-term constant manner (we refer to this as a steady state), the CCD matches the depth of the "snow line" that marks the transition between sediments containing carbonate above it, and sediments without carbonate below it. This is similar to the snow line on mountains, where snow is present above the line and absent below the line. The CCD exists because CO_2 pressure goes up with water pressure, because the cold water that fills the deep sea can hold a lot of CO_2 to begin with, and because CO_2 in the deep sea increases with the age of the deep water (where "age" is a measure of how much organic matter decomposition has taken place in the water mass). Thus, modern Pacific deep waters are more corrosive than Atlantic deep waters, because Pacific deep water is much older than Atlantic deep water because of the dominant deep-water circulation patterns (see figure 7). The more corrosive the deep water is, the shallower the CCD will sit, and the more of the deep sea will be carbonate-free. In consequence, the CCD resides at about 4.2 to 4.5 kilometers' depth in the Pacific, and at about 5.0 kilometers' depth in the Atlantic.

However, things are not always about contrasts between shallow and deep water, as they are with the biological pump. If a period of intense weathering causes enhanced transport of Ca^{2+} and HCO_3^- into the ocean, then the carbonate reaction everywhere is nudged more strongly toward the carbonate-formation side. As a result, the CCD deepens, and there is improved carbonate preservation in the deep sea while carbonate production also increases in shallower waters. Effectively, the entire ocean shifts to a more carbonate-friendly environment in such a case. Several other processes come into play for controlling CCD depth changes, especially because of changes in the ratio of organic carbon versus carbonate sinking into the deep sea (known as the rain ratio). But these represent rather technical issues that go beyond the scope of our story, which concerns the main changes in organic- and inorganic-carbon-cycle interactions through Earth's history between the Cretan, Neritan, and Strangelove oceans.

The buffering of CO_2 fluctuations by well-established carbonate compensation in the world ocean was, importantly, improved by the advent of planktonic calcifiers in the (modern-style) Cretan oceans of the last 252 million years, and most notably of the last 170 million years. It has been a key factor in stabilizing shorter-term CO_2 and climate fluctuations over time scales of many thousands of years. Today, the carbonate burial into marine sediments that compensates for weathering influxes of Ca^{2+} and HCO_3^- into the ocean is divided about equally between deep-water and shallow-water zones. During ice ages, global sea-level lowering of 100 meters or more caused the area of shallow-water zones to dwindle to a quarter of its previous size, and deep-water burial became more important. This "resilience" of the carbonate-burial processes did not exist during Neritan ocean times, as there were no planktonic calcifiers to maintain deep-water burial (see below). Ironically, the Cretan oceans' widespread carbonate burial in deep-water sediments, and their ensuing subduction and decarbonation, promoted large plate-tectonically driven CO_2 and climate variations over longer time scales of millions of years.

The Neritan oceans, between the Cambrian explosion of 542 million years ago and the end-Permian extinction of 252.3 million years ago, also contained many organisms with carbonate skeletal parts, but they were restricted to the seafloor, especially in shallow-water regions. As a result, the amount of biological carbonate burial in the deep sea was low, in contrast to abundant shallow-water carbonate formation; shallow-water-platform carbonates were more strongly developed in Neritan oceans than in Cretan oceans. During some times of low sea level within the Neritan period, however, the shallow regions became so restricted in size that biological carbonate burial throughout the ocean dropped precariously, at which times widespread abiotic (chemical)—though likely microbially assisted—carbonate deposition developed, similar to conditions that are known from Strangelove oceans (below). Because of the lower deep-sea carbonate burial in

Neritan oceans than in the modern Cretan oceans, there was a much less effective deep-sea carbonate buffer. This allowed more intense CO_2 and climate swings in association with sea-level changes during Neritan times than during Cretan times.

Incidentally, there may have been no planktonic calcifiers in the Neritan oceans, but there were free-swimming carbonate-shelled organisms such as ammonoids and belemnoids, which appeared at about 400 million years ago (they continued until about 66 million years ago) (see figure 2). Like most carbonate producers of Neritan times, these mostly lived in warm shallow-water regions. Ammonoids were molluscs with mostly spiral-shaped carbonate shells that resembled modern *Nautilus* but were more closely related to modern octopus, squid, and cuttlefish. Throughout their existence, ammonoids underwent distinct evolutionary shell changes, which underpin their extensive use as index fossils to evaluate geologic time zones in marine sediments. Belemnoids were squid-like in appearance, but had 10 arms of equal length and had hard, often bullet-shaped, internal carbonate skeletons.

Finally, then, some thoughts about the carbonate Strangelove oceans that dominated from very early times until the Cambrian explosion. These were very alien to us in that they were dominated by abiotic carbonate precipitation, including signs of distinct microbial influences from calcified cyanobacteria; the cap carbonates directly after snowball-Earth episodes are good examples. Chemically, abiotic carbonate precipitation would have been focused on nucleation and crystal growth on organic and inorganic surfaces in warm, shallow waters. Strangelove oceans allowed very limited buffering of the global carbon cycle through deep-water carbonate compensation for any shift in sea level. This is thought to have "allowed" the development of extreme CO_2 responses, with intense swings between extreme climate states, in the times before the Cambrian explosion, including snowball Earth.

In simplified terms, the development from Strangelove oceans to Neritan oceans and then to Cretan oceans may be viewed as a stepwise introduction of more and more carbonate-compensation-based buffering of climate swings, including sea-level fluctuations. This progressive intensification of the buffering efficiency caused increased dampening of climate cycles through Earth's history. Such cycles were poorly dampened in Strangelove oceans without much biological calcification, better dampened in Neritan oceans with development of widespread calcification at the seafloor in shallow regions, and strongly dampened in Cretan oceans with well-developed deep-water carbonate burial due to planktonic calcifiers in conjunction with well-developed shallow-water carbonate burial.

If you found all this talk about organic carbon, carbonate, and their interactions very confusing, then I welcome you to my world. Indeed, the interactions between the organic and inorganic sides of the carbon cycle are challenging for most who do not work on them every day. But they are critical to many aspects of the development of the oceans, climate, and life. To ensure that all—often counterintuitive—responses are properly captured, calculations or modeling exercises are essential. For further reading on this bewildering topic, I recommend starting with the studies by Andy Ridgwell (from the United Kingdom, now in California) and Richard Zeebe (from Germany, now in Hawaii) that are listed in the bibliographies of chapters 4 and 5 and to pursue the references given therein. But make sure to have some headache medicine handy; I for one always need some.

*****

To wrap up this chapter, we can conclude that the rather abstract aspects and implications of snowball Earth and the explosions of life brilliantly illustrate the intricately interlocking nature of the various biological and geologic processes within the Earth system. The development of many feedbacks, checks, and balances within this system determined why conditions on Earth could always remain suitable for organic life, and why life eventually underwent a step change that allowed colonization of Earth by complex multicellular plants and animals. We have seen how the oxygenation history relates to early developments of life and in particular to photosynthetic developments and the rise to prominence of sponges, how release of decay products (DMS, nutrients, carbon itself) drove or accelerated dramatic climate changes, and how Earth's plate tectonics was essential for long-term CO_2 reduction leading up to snowball-Earth episodes as well as CO_2 buildups that terminated these episodes. And finally, we have learned how modern-type nutrient and carbon cycles were established that created suitable conditions for nourishing further developments of life, including the progressive stabilization of climatic fluctuations.

Next, we will see the carbon cycle take central stage in subsequent notable events in Earth's ocean and climate history.

CHAPTER 5

OCEANS ON ACID

The first vertebrate animals appeared in the oceans at around 525 million years ago, within the Cambrian explosion. Jawless fish similar to lampreys and hagfish also appeared in the fossil record of the Cambrian period (see figure 2). Some of the best-preserved fossils are the conodonts, bony toothlike elements belonging to fish that in shape resembled modern lampreys or eels.

The conodont fish survived for a long time, until a big extinction event at around 201 million years ago. Meanwhile, the first fish with jaws, the acanthodians, had appeared shortly after a major extinction event 444 million years ago, and not long after that the first placoderms—armored jawed fish—had shown up too, as did the cartilaginous fish (sharks, rays, and skates) and bony fish (most other fish). A period known as the Devonian ensued from 419 to 359 million years ago, which has become known as the "age of fish" because fish were the dominant vertebrates of the time (see figure 1). As we saw before, the geologically important ammonoids appeared at around 400 million years ago, as did the belemnoids (see figure 2).

The placoderms, some of which were enormous at 8 to 11 meters in length, vanished in an extinction event at the end of the Devonian period at around 359 million years ago. This allowed rapid expansion of the cartilaginous and bony fish, which have dominated fish faunas ever since. Acanthodians survived until their demise as one of the victims of a major extinction event 252.3 million years ago, when disaster truly struck. At that time, marking the end of the Permian period, the world experienced its worst mass-extinction event.

As an aside, paleontologists recognize more than 20 notable extinction events through Earth's history. The 5 greatest extinctions, in chronological order, were the double-spiked Ordovician-Silurian extinction of 450–440 million years ago, the prolonged series of Late Devonian extinctions of about 375–359 million years ago, the end-Permian extinction at around 252.3 million years ago, the end-Triassic extinction of 201 million years ago, and the infamous Cretaceous-Paleogene boundary extinction of 66 million years ago that was formerly known as the Cretaceous-Tertiary boundary (see figure 1). The ammonoids and belemnoids were remarkable survivors through several of these extinction events, albeit in small numbers and undergoing major species "resets" through each event. They were eventually terminated in the Cretaceous-Paleogene boundary extinction event.

The end-Permian extinction was the worst of all extinction events. About 96% of marine life and 70% of land-based life became extinct. In the oceans, the extinction especially affected marine organisms with calcium carbonate skeletons, including the reef builders of the time. The interval over which the end-Permian extinction happened (in two phases) was about 60,000 years long in the oceans, but the main impact on land plants took a few hundred thousand years. Recovery started in earnest at about 5 million years after the event, and it took up to 30 million years for biodiversity to fully recover. Even then, the individual species and groups that had gone extinct did not return, but were replaced by entirely new ones. As discussed, a new mode of ocean carbonate formation started to gain ground after this extinction event, centered on the planktonic calcifiers (see figure 2). As with all major extinction events, the end-Permian event represented a major "reboot" of the Earth system. Although the planet and life in general survived, life as it was known before the event was obliterated and replaced.

The extinction event has been related to a very fast addition of a large quantity of external carbon into the atmosphere-ocean-biosphere system, where the term "external" means from a reservoir outside the active, or internal, atmosphere-ocean-biosphere system. This led to a triple whammy of intense global warming, ocean anoxia (no-oxygen conditions), and then severe ocean acidification in the main phase of the extinction.

Next, we explore what ocean acidification is, and how it is recognized in the geologic record. We will then use this to shed some light on the end-Permian extinction as well as other, more recent, events of the oceans on acid.

ABOUT ACIDIFICATION

In general, the oceans acidify when increasing levels of dissolved CO_2 make waters more corrosive to carbonate. But the term "ocean acidification" is not commonly used when talking about processes associated with the biological pump that drive conditions in surface and deep waters in opposite directions. Instead, we use the term more specifically for the impacts of a large and fast input of external carbon, which drives conditions everywhere throughout the oceans in the same direction.

Let's look at it in more detail. We start with what we know best; namely, how acidification manifests itself in the modern-style Cretan oceans that include planktonic calcifiers (the last 252 million years, and especially the last 170 million years). Thereafter, we consider differences that may be expected in the Neritan oceans, which contained no planktonic calcifiers and thus had very limited deep-water carbonate burial (542 to 252.3 million years ago). By discussing things in this order, we usefully set the scene for evaluating another, younger, and better-documented ocean acidification event that dates to 56 million years ago (in Cretan ocean conditions), in comparison to the end-Permian event.

In the modern-style Cretan oceans, the term ocean acidification refers to changes that are faster than the buffering action of deep-sea carbonate compensation, which in practice means over time scales of less than 10,000 years. Slower, long-term changes cause steady-state, matching shifts in the depth of the CCD and the carbonate snow line. But fast changes can temporarily break this association; fast addition of external carbon acidifies the ocean, and the CCD rapidly shifts to a shallower depth. This means that it shifts temporarily to a location above the carbonate snow line because it takes time to dissolve carbonate in the sediments that have now ended up below the

CCD. Next, the slow dissolution process gradually brings the snow line up until it again matches the CCD. This takes many thousands of years. Eventually, a new balance is reached because dissolution of carbonates lowers the CO_2 pressure in the deep sea.

The external carbon mentioned above typically comes in the form of CO_2 or CH_4 from rocks and sediments of the lithosphere. Volcanic outgassing and—in the past few centuries—fossil-fuel extraction and combustion release carbon that has been locked away for millions to hundreds of millions of years. Gas-hydrate breakdown releases external carbon that has been locked away for tens or hundreds of thousands of years. These external-carbon injections upset the balance of the active system's carbon cycling, which activates—to varying degrees—two key processes of carbon removal until a new balance is achieved between carbon input and output.

The first process, operational throughout Earth's history after the appearance of life, involves the burial of organic carbon in sediments—on land, in the oceans, or both—and is dealt with elsewhere in this book. Carbon deposition and burial in sediments can offset major external-carbon injections. But it is a slow process that requires many tens of thousands of years or more.

The second response to external-carbon injection is ocean acidification, which in turn becomes gradually decreased through carbonate compensation in Cretan-ocean times (in Neritan-ocean times, without well-developed deep-sea carbonate compensation, reduction of acidification spikes would have been less straightforward and effective).

During strong ocean acidification, the reducing carbonate saturation state throughout the oceans stresses organisms that deposit carbonate for their skeletal parts. This affects the shape, mass, and size of the skeletal parts. Such variations have been seen in microfossil records through past ocean acidification events.

In today's (Cretan-style) world ocean, the concentration of H^+ ions—and thus the pH—is strongly buffered by reactions between the different compounds of the complex ocean chemistry, and especially by the carbonate chemistry. The pH of the surface ocean had settled around a value of 8.2 before the fast injection of external CO_2

that started with the industrial revolution. This value represented a long-term balance between the controlling processes that had been established over a more than 6000-year stable phase of the current interglacial period. Since the onset of the industrial revolution, surface-ocean pH has been decreasing, which marks notable acidification. Today, the pH is around 8.1. Although a drop of 0.1 pH units may seem small, pH is a logarithmic function of the acidity. Thus, a drop of 0.1 represents a 25% increase in acidity over the past two centuries. If we continue with our emissions as we are today, we will end up with a surface-ocean pH of about 7.8 or 7.7 by the end of this century. Past acidification events can be used to evaluate whether, and how, this may affect marine life.

All organisms thrive on chemical reactions that are very sensitive to the degree of acidity of the environment and their body fluids. Humans, for instance, have a body pH of about 7.4. A drop by only 0.2 can cause serious health issues, such as seizures, coma, and death. In marine organisms, similarly small changes in pH, governed strongly by that of their watery environment, especially affect reproduction and growth. For example, these organisms will struggle to dissociate bicarbonate ions to free up the carbonate ions they need, under conditions of increasing H^+ concentrations. In consequence, they will struggle, or at least have to expend more energy, to form carbonate for their skeletal parts. Other impacts of increasing acidity include difficulties for coral larvae in finding new places to settle, trouble for mussels in forming strong threads that hold them in place, inhibition of early shell formation in young oysters, and issues for larval sea urchins in digesting food. Also, some studies found that carbonate-forming phytoplankton algae develop problems building their skeletal parts, although other studies found that they eventually manage to adapt. Moreover, fish-blood pH is in equilibrium with the water they inhabit, and—like humans—fish suffer very serious health risks for a pH drop of only about 0.2. Trying to restore its pH when it drops costs a fish much energy; this stifles its growth and slows its mobility.

In Neritan-ocean times with a weaker potential for deep-sea carbonate compensation, large and fast external-carbon releases have also occurred (we will discuss one example below, for the end-Permian extinction event). Because of relatively low carbonate deposition in these times, the ocean waters are thought to have been strongly supersaturated, especially warm, shallower waters. This gave the water itself a considerable buffering capacity to CO_2 invasion due to external-carbon releases, although that capacity would have been weaker than the buffering capacity of the Cretan ocean's deep-sea carbonates.

Once the buffering capacity of the waters was overcome by fast and large external-carbon injection, ocean acidification would ensue. At the same time, atmospheric CO_2 levels would be very high. High atmospheric CO_2 levels, high temperatures during intense greenhouse conditions, an abundance of fresh volcanic rocks, and an increasing carbonate saturation state in the oceans due to collapsed carbonate deposition in acidified shallow waters all would have helped to create conditions that favored gradual net drawdown of the externally injected carbon levels. This would be assisted by any increases in the rate of organic-carbon burial if widespread anoxia developed. These processes of system recovery after a fast external-carbon injection in Neritan-ocean times may have worked over similar, or more likely longer, time scales than those in Cretan-ocean times, given that Cretan oceans possessed all the same response processes plus the additional carbonate-compensation process.

*****

Despite its well-known importance for marine life, ocean acidification has proven hard to precisely reconstruct through past ocean history. In the remainder of this section, I outline three main lines of evidence that have been explored—namely, boron-based methods, CCD tracking, and microfossil-fragmentation indices. Combined, they give pretty decent information for the Cretan-ocean times with planktonic calcifiers. For the earlier Neritan-ocean times without planktonic calcifiers and therefore with limited deep-water carbonate burial, however, the latter two methods do not apply. So I first evaluate the boron-based methods, which show promise but require further work to improve confidence. Thereafter, for completeness and to lay foundations for discussion of younger acidification events, I outline the CCD-tracking and microfossil-fragmentation methods that apply only under Cretan-ocean conditions.

The boron-based methods are fast gaining prominence for reconstructing past surface-water CO_2 levels and deep-water carbonate chemistry. They are exciting to researchers because they show good promise for delivering more detailed, usefully quantified reconstructions. However, pitfalls and uncertainties remain, especially concerning the validity of the data in view of changes in ocean chemistry through time, and of alteration of the original chemical signatures of the measured carbonates after their deposition. The boron applications fall into two categories. Boron isotope ratios are used for reconstructing surface-water CO_2 levels. Boron-to-calcium (B/Ca) ratios are used for reconstructing deep-water carbonate chemistry; boron isotope ratios are not commonly used for this because sample-material availability is too limited.

Boron isotope studies for Cretan-ocean times analyze boron isotope ratios in fossil carbonate shells of surface-dwelling microzooplankton (planktonic calcifiers) that are typically found in seafloor-sediment cores. The method then relates shifts in the isotope ratio to changes in ocean-water pH. Next, a series of calculations uses the reconstructed pH to give the oceanic CO_2 level, which is related via gas-exchange calculations to the atmospheric CO_2 level. Comparison with other oceanic and terrestrial CO_2 proxies shows reasonable agreement about large-scale CO_2 variability over millions of years. There is still plenty of scope in the finer details for the researchers to engage in a good deal of bickering—and they enthusiastically do so. With caution, researchers have started to propose results from the boron isotope method for Neritan-ocean times, and in particular for the end-Permian extinction event. Reconstructions for these times cannot rely on the remains of planktonic calcifiers for analysis, and instead use bulk samples of shallow marine open-water carbonates. This provides cause for an additional layer of disagreement and argumentation, and the jury remains out on exactly how good the reconstructions may be.

B/Ca ratios are analyzed using carbonate microfossils of seafloor-inhabiting unicellular organisms called benthic foraminifera. The ratio reflects changes in past carbonate ion (CO_3^{2-}) concentrations in the deep sea, and calculations relate these to the amount of CO_2 dissolved in the water. I have not yet seen this method applied in geologic intervals older than a few hundred thousand years, but I may have missed a study or two. In any case, the B/Ca method is not yet relevant to the problem addressed in this chapter.

The two other main methods of reconstructing past ocean acidification changes strictly apply only to Cretan oceans. One method relies on tracking changes in the depth of the CCD, using well-dated drill cores from different water depths. The other uses the state of preservation of carbonate microfossils to assess carbonate dissolution in the deep sea.

Reconstructions of CCD changes through geologic time were pioneered in the early 1970s by the German and Dutch geoscientists Wolfgang Berger and Tjeerd van Andel, who were based for much of their careers in the United States (incidentally, van Andel made several deep-sea dives with the US submersible *Alvin*, including one in 1977 offshore Ecuador, during which the team made the first direct visual observations of hydrothermal vents on the seafloor at more than 3000 meters' depth). Following the pioneering work, many reconstructions followed, but mostly in rather coarse time steps. A detailed study led by the German paleoceanographer Heiko Pälike, my former neighbor at the University of Southampton in the United Kingdom, took things to a new level in 2012. It revealed not only a long-term 1.6-kilometer deepening of the equatorial Pacific CCD over the past 55 million years, but also a previously unknown series of superimposed shorter fluctuations over depth ranges of up to 600 meters. Calculations with an Earth-system model attributed these large changes to a combination of major changes in weathering and in the organic carbon cycle (biological pump), which were slow enough to drive gradual steady-state shifts.

In steady-state shifts, the CCD and the snow line remain closely together in response to gradual changes in the net input of external carbon. Such gradual changes may, for example, occur in response to a long-term increase in volcanic CO_2 outgassing, but can also take place because of a long-term decrease in weathering or in carbon burial, which causes a decrease in CO_2 removal from the atmosphere-ocean-biosphere system (for constant inputs from outgassing, a decrease in CO_2 removal gives a net increase in external carbon). In reverse, a period of more intense weathering or carbon burial drives a net external-carbon decrease when outgassing remains constant, which results in a CCD shift to greater depths.

CCD shifts due to net external-carbon changes have a rather global signature; they won't be exactly the same everywhere, but at least the movements will be (virtually) everywhere in the same direction. CCD shifts due to changes in the operation of the biological pump will be different in magnitude and direction in different places, and likely will be associated with fossil evidence of change in the ecosystem that drives the biological pump. But things get really interesting when external-carbon injections are fast and large enough to temporarily overwhelm carbonate compensation. In those cases, an ocean acidification event develops rapidly, with temporary separation between the CCD and the marine snow line, followed by slow recovery.

In warm surface waters, which are normally saturated for carbonate, dissolution will not normally become evident. But carbonate formation may become more difficult as the degree of saturation drops, so that calcification rates become affected in many algae and zooplankton. In the deep sea, however, a large and fast external-carbon injection moves the CCD up by many hundreds of meters. This upward shift causes carbonate-corrosive waters (below the CCD) to bathe deep-sea sediments that contain an awful lot of carbonate owing to accumulation over time of the skeletal parts of dead planktonic calcifiers. Sedimentary carbonate then starts to dissolve. This carbonate-compensation process triggered by the rapid upward CCD shift is an effective way of offsetting the impacts of the initial external-carbon injection. In other words, carbonate compensation in Cretan-ocean times (the last 252 million years) is nature's perfect way of buffering an acidification event. It's just that the time scales are different. The initial fast net external-carbon input causes harmful acidification when the carbon input temporarily overwhelms the carbonate-compensation capacity on time scales of up to 10,000 years. Thereafter, the slower carbonate-compensation process restores the system to a new balance, which can take up to a few hundred thousand years. We will explore a particularly well-documented example of this sequence in the Acidification in Action section later in this chapter, when we discuss the Paleocene-Eocene Thermal Maximum (PETM) of 56 million years ago.

The second method of reconstructing past ocean acidification changes that applies only in Cretan-ocean times employs the state of preservation of carbonate microfossils in the deep sea. As we approach the CCD from above, carbonate microfossils in the sediment show increasing signs of partial dissolution and fragmentation, which results from weaknesses in the shells caused by the partial dissolution. Researchers have been making fragmentation indices, shell-weight indices, and combinations of these to evaluate dissolution state and to infer from that the deep-sea carbonate chemistry. This can help in resolving carbonate chemistry changes above the CCD, including changes leading into, and coming out of, ocean acidification events. Note that shell weight is also affected by the carbonate saturation state of the surface waters in which the algae or zooplankton were living. Overall, preservation indices are an interesting concept, but the information obtained from them is open to a rather wide range of interpretation, and new (notably boron-based) approaches are rapidly taking over.

One thing stands out from all this work on trying to reconstruct ocean acidification changes: we're just beginning to understand how acidification impacts life. This includes the impacts of human-induced external-carbon emissions. There are serious concerns because any detrimental impact near the base of the marine food web will reverberate right through it, affecting the ecosystem as a whole. As a result, it is important to investigate the impacts of past acidification events, while carefully accounting for the type of ocean in which they occurred, be it Cretan, Neritan, or Strangelove. The next section evaluates signals through the end-Permian extinction of 252.3 million years ago that occurred in a Neritan-style ocean, and compares this with the much younger PETM acidification event of 56 million years ago that occurred in a Cretan-style ocean.

ACIDIFICATION IN ACTION

Our first case history of ocean acidification concerns the end-Permian extinction of 252.3 million years ago, which happened under Neritan ocean conditions. The end-Permian extinction happened over a time span of about 60,000 years in the oceans. Carbon isotopes in the oceans record a shift 10,000 years before the extinction to increased carbon-12 relative to carbon-13, which is a strong indicator of external-carbon injection. Outgassing during emplacement of the Siberian Traps flood basalts (see figure 13) is a primary suspect. These eruptions crossed thick beds of coal and carbonates, which caused elevated releases of CO_2, as well as sulfur, and which were critical contributors to the global carbon isotope signals seen through the extinction-event sequence.

Reconstructions suggest that a first major phase of external-carbon injection led to strong warming of the order of 8°C, and to widespread oxygen consumption in the oceans that kicked off the extinctions in about 10,000 years. Biomarkers (characteristic organic molecules) of a group of organisms called Chlorobiaceae, or green sulfur bacteria, have been found in abundance in sediments from the event and indicate that large portions of the oceans became anoxic. And a particularly rapid expansion of ocean anoxia took place shortly before the onset of extinctions. Eukaryotes took a big hit during the extinction event, and prokaryotes rapidly reoccupied the vacated ecological niches. In more ways than one, therefore, one might surmise that the oceans experienced a sort of relapse to the anoxic, prokaryote-dominated conditions that existed before the rise of eukaryotic algae about 800 million years ago.

Some 50,000 years after the carbon isotope shift, a sharp acidification event took place. Boron isotope data suggest a dramatic surface-water pH drop of 0.6 to 0.7 pH units within about 10,000 years. This acidification event involved an abrupt second external-carbon release, whose rate of emission may have been close to that of current human-induced emissions. But note that the rate of recorded acidification of up to 0.7 pH units in 10,000 years was still much slower than the projected drop of up to 0.5 pH units in about 250 years of the human-caused acidification by the year 2100. Regardless, the end-Permian acidification clearly reflects that the external-carbon emissions had overwhelmed what buffering capacity the ocean may have had at the time. The resultant acidification spike heralded the deadliest phase of the extinction period. All existing reef-building organisms died out in the oceans. The carbonate-forming corals of today all stem from developments that started directly after the end-Permian extinction event. Ammonoids and belemnoids survived, but only barely so.

While all this may sound pretty absolute, in reality there remains plenty of healthy debate about both the nature of the evidence and its interpretation in terms of a sequence of events. Some have attributed the sharp second phase of the extinction event to an asteroid or comet impact, but evidence for this remains thin. As it stands, the sequence of events described above seems to be most in line with the available evidence. Unfortunately, though, this evidence will always remain rather limited. Dating to 252 million years ago, the end of the Permian is notably older than the oldest ocean crust. This means that we cannot drill to any unaltered sediments of this period in the seabed. And that, in turn, limits evidence collection to sediments that have become caught up in uplift and mountain formation, and which commonly have become considerably altered and/or mangled. Pristine deposits exist, but they are rare, and this limits our capacity to achieve sufficient detail in reconstructions through the end-Permian event.

Still, the end-Permian mother of all extinctions is associated with a large external-carbon injection at rates approaching those of the current human-induced emissions. We are therefore well advised to figure out ways of unraveling its secrets as soon as possible, lest we inadvertently slide down the slippery slope into a similar mess in the near future. The limited availability of material for the end-Permian then calls for creative alternative approaches for achieving a deeper understanding of acidification impacts. The typical way of tackling such a conundrum is to compare and contrast the event with information from more recent—albeit less intense—events, which can be sampled and investigated in much greater detail.

The remainder of this section follows that philosophy. We temporarily break away from the book's essentially chronological assessment of events to look at a less severe, but more recent and much better documented, acidification event: the Paleocene-Eocene Thermal Maximum (PETM) of 56 million years ago.

*****

The PETM acidification event took place in a Cretan-style ocean. As such, the PETM offers more directly relevant insights into the potential impacts of today's ocean acidification due to human-induced carbon emissions, given that both occurred under Cretan ocean conditions.

From about the time of the end-Permian extinction, high-CO_2 greenhouse conditions had started to build up. These conditions peaked at levels of around 2000 ppm and according to some even up to 4000 ppm, at some time between 250 and 150 million years ago. Different reconstruction methods disagree about the exact age and size of the peak. Starting 150 million years ago, an overall, but considerably variable, cooling took place lasting until the present; that is, from high-CO_2 greenhouse conditions of the Mesozoic era, which we will discuss in chapter 6 (252.3 to 66 million years ago), through to the low-CO_2 icehouse conditions of the past few million years.

The PETM began about 56 million years ago, when CO_2 levels were still very high, between best estimates of about 650 and 1050 ppm. The PETM itself is marked by a remarkably sharp shift in carbon isotope data due to an increase in the proportion of carbon-12 over carbon-13. As with the end-Permian event, this is key evidence of an external-carbon input. CO_2 levels shot up to best-estimate values between about 1400 and 3350 ppm, and global average temperatures spiked up by about 6°C (see figure 14). The event lasted less than 200,000 years, but it was very asymmetrical: less than 20,000 years were needed to go from first signals to the peak (including a very rapid initial event within a few thousand years, followed by more gradual continuation), and the remainder of the time concerned recovery from the disturbance.

The PETM was a hyperthermal, an interval of extremely high temperatures and low latitudinal sea-surface temperature gradients, and developed from a background state that had very high temperatures already. A few other hyperthermals occurred around the time of the PETM, although these were considerably weaker. A notable one dates to about 53.7 million years ago. It is called ETM-2 (Eocene Thermal Maximum 2), but it is more affectionately known as ELMO. On board ship, the name ELMO was given in reference to the *Sesame Street* character, because the clay layer in the sediment cores was red. It was then explained as "Eocene layer of mysterious origin" (eight weeks on a research ship does somewhat interfere with one's state of mind). Somehow the joke managed to escape the confines of the vessel.

The PETM warming pushed global temperatures up by about 6°C relative to conditions before the event, which were warm already at estimated global average values up to 25°C, which is up to 10°C higher than values before the industrial revolution (see figure 14). The Arctic was ice-free, with summer temperatures that—according to organic geochemical measurements—rose from about 17°C before the event to about 23°C during the event. There was a remarkably small temperature contrast between the equator and the poles. Although this notion may be hard to understand, it relies on many types of evidence. For example, there were rich forests at high northern latitudes and even tropical palm forests at mid and high latitudes in North America and Europe. Tropical algae are found in sediment cores from high latitudes. Forests covered large parts of Antarctica. And even the general warm conditions of the Eocene, superimposed upon which the PETM hyperthermal developed, already sufficed to support fossil crocodilians on Ellesmere Island, at 78° N. To all accounts, the PETM was a truly exceptional warming event on top of an already warm and moist climate state.

Estimates for the PETM's external-carbon addition range between 2000 and 7000 gigatonnes (Gt; or GtC for gigatonnes of carbon, where one gigatonne is one billion metric tons, or one thousand billion kilograms). Some of the warming of the event started before the major carbon addition and may have caused the carbon release. For scale, human-induced carbon emissions since the industrial revolution have been about 420 GtC, and are currently rising faster than ever, by almost 10 GtC each year (or almost 40 Gt CO_2). Although smaller in total volume than the external-carbon injection that triggered the PETM, the human-induced emissions since the industrial revolution have been 5 to 30 times faster. Similarly, at 6°C the PETM warming was larger, and it took less than 10,000 years—the most recent estimates suggest 4000 years. That still makes it almost five times slower than the warming of about 1°C since the industrial revolution.

So, as a brief aside, it appears that current human-induced climate change is moving much faster than the dramatic PETM event. Cleanup of the external-carbon injections by natural mechanisms, however, will take similar time scales for the current changes as it did for the PETM—almost 200,000 years. If we want a faster cleanup, then we will need to invent effective ways of helping nature with it, similar to the highly efficient ways in which we have "helped" pumping external carbon into the system. This will require global large-scale application of technological solutions to capture CO_2 at source and to remove CO_2 from the atmosphere, on a routine round-the-clock basis, but such technology is—at best—still in its infancy today. I reemphasize: the geologic record (the PETM being a case in point) clearly shows us that nature alone cannot remove our CO_2 fast enough. Left to its own devices, nature would need a few hundred thousand years to do the job.

The large and, by natural standards, fast external-carbon injection of the PETM had a dramatic impact on the oceans, mainly because the injection was fast enough to temporarily overwhelm the slow carbonate-compensation process (given that modern emissions are rising much faster, we may expect the same in our future). The PETM's external-carbon impacts were of a similar nature to those of the end-Permian, albeit less extreme. They included strong warming, widespread development of low-oxygen conditions in the oceans, and remarkable ocean acidification.

Up to 50% of unicellular bottom-dwelling carbonate-forming foraminifera went extinct in the deep sea during the PETM, but we still do not yet fully understand why or how. Suggested reasons include the lowered deep-sea oxygen levels of the event, as well as a more than 5°C deep-sea temperature rise (warming speeds up metabolism, and thus the biological use of oxygen in the deep sea—metabolic rates roughly double for every 10°C rise in temperature), widespread ocean acidification, and changes in the food supply to the deep sea due to surface-water productivity changes. The responses of surface-water organisms to the event's ocean acidification impacts were a mixed bag; there are reports of an increase in heavily calcified algae (a potential protection strategy) and in weakly calcified zooplankton. Other reports mention extensive deformations.

Ocean acidification during the PETM especially affected the Atlantic Ocean, and to a more limited extent also the Pacific Ocean. An Ocean Drilling Program expedition in the early 2000s, led by the paleoceanographers Jim Zachos from the United States and Dick Kroon from the Netherlands (now in Scotland), recovered a transect of sediment cores from the Walvis Ridge, offshore southwestern Africa. The transect spans about two kilometers of water depth, and the PETM is expressed as a very clear carbonate-dissolution event over the entire transect. This relates to a CCD shift to shallower depths over two kilometers, within about 10,000 years. It then took well over 100,000 years for the system to revert to more normal conditions. In the Pacific Ocean, the CCD only moved to shallower depths by a few hundred meters. So a large part of the Atlantic signal is region specific, and the global average CCD shift is more modest. But don't let this give you the wrong impression; a global average CCD shift by a few hundreds of meters still represents an enormous change in the carbon cycle.

External CO_2 has relatively carbon-12 dominated carbon isotope values, but external CH_4 has even lighter carbon isotope values. To explain a global carbon isotope shift to more carbon-12 dominated values, we therefore need very large quantities of external CO_2, or relatively smaller quantities of CH_4. Note also that CH_4 can be produced in nature along different pathways. We won't dwell on that, other than noting that biogenic CH_4 is isotopically much lighter than thermogenic CH_4, so that a release of biogenic CH_4 would cause a stronger global carbon isotope shift than a similar release of thermogenic CH_4. Regardless, the total amount of external-carbon input, no matter whether it is as CO_2 or CH_4, together with the rapidity at which it is injected, determines the global average CCD response. Taking these pieces of information together, many researchers have attempted to calculate what happened.

The American geochemist Jerry Dickens determined in 1995 that the carbon isotope shift suggests an injection of 1000 to 2000 GtC's worth of CH_4 from gas-hydrate breakdown. More recent work has also taken into account the CCD changes, and suggests closer to 3000 GtC. But not everyone is convinced that all the injected carbon came in the form of isotopically very light CH_4. Releases of isotopically heavier CH_4 and CO_2 have been suggested; for example, from permafrost melting on the Antarctic continent at that time. Such scenarios require a larger volume of external-carbon input to satisfy the isotopic observations (up to 7000 GtC), but then issues arise in matching them with the CCD changes, as these would need to be even larger than observed.

As ever, a lot of debate remains about the exact mechanisms, sources, and quantities, but it seems beyond doubt that more than

2000 Gt of external carbon was injected. A considerable portion of it may have come in the form of CH_4. The additional signals of low-oxygen conditions at many depths levels in the oceans have been explained mainly in terms of poor deep-water circulation (and thus oxygenation) in response to the rapid warming, with additional impact from the oxidation of external carbon in the system.

Signs of reduced deep-water oxygenation for the PETM extend its similarity (in a less intense form) with the end-Permian acidification event. This calls for a discussion of deep-sea oxygenation changes. However, deep-sea oxygenation crises occurred again during the Mesozoic, and very prominently so. The Mesozoic is the next stop on our journey through the history of the oceans, and discussion of deep-sea oxygenation crises will therefore follow in chapter 6.

*****

To recap, this chapter has explained what ocean acidification is, including differences in its nature in Cretan and Neritan oceans. We evaluated how ocean acidification is caused by large and fast injections of external carbon into the climate system, and by which means we can document and distinguish both steady-state changes and ocean acidification. We discussed a major acidification event during the end-Permian extinction under Neritan ocean conditions, and we compared and contrasted it with the PETM acidification event under Cretan ocean conditions. Finally, we saw that the broader phenomena of both the PETM and the end-Permian extinction also included deep-sea oxygenation reduction, or full-blown anoxia. This brings us neatly to the remit of the next chapter, which includes a more detailed discussion of such conditions.

CHAPTER 6

THE AGE OF REPTILES

After our digression into the PETM, we now retrace our steps to where we left the main thread of the story: the end of the Permian period, a spell of greenhouse conditions that saw the extinction of about 96% of marine life. At that time, CO_2 levels were rising fast, and the end-Permian extinction itself coincided with an even more accelerated rise. This brought Earth from Late Carboniferous–Permian icehouse conditions into a very long spell of greenhouse conditions, which effectively extended right across the Mesozoic era (252.3 to 66 million years ago) and beyond, until the end of the Eocene epoch at 33.9 million years ago. In all, the end-Permian event heralded a period of greenhouse conditions that lasted almost 220 million years.

This long greenhouse interval was a period of great experimentation and success with new life-forms and life strategies. In the ocean, the ammonoids and belemnoids had almost gone extinct in the end-Permian event, but the scant survivors rapidly radiated into great numbers and species diversity. Also, calcifying algae and zooplankton appeared on the scene—the planktonic calcifiers. These would gain importance until a major revolution in carbonate cycling within the oceans was achieved at around 170 million years ago, at which time the planktonic calcifiers reached a dominant presence; some call this the Mid-Mesozoic Revolution.

Through the Mesozoic era, reptiles would come to rule the planet. Reptiles had first emerged in the Late Carboniferous, between about 320 and 310 million years ago (see figure 2). Within a few million years after the end-Permian extinction, some reptiles transitioned from land into the sea, as documented in recent findings of the likely amphibious "transitional" *Cartorhynchus*. Through the Mesozoic, Earth would see the rise and demise of dinosaurs on land, as well as of giant predatory reptiles such as mosasaurs, ichthyosaurs, pliosaurs, and plesiosaurs in the oceans. The Mesozoic era ended with the infamous Cretaceous-Paleogene (formerly known as Cretaceous-Tertiary) boundary extinction, which terminated the age of reptile domination and opened the ecosystems up for rapid expansion of mammals, fish, and birds.

During the warm and equable Mesozoic conditions, there was intense reef formation throughout the oceans, and especially along the low-latitude margins of the Paleo-Tethys and (Neo-)Tethys Oceans. It's useful to start our discussions of ocean and climate changes with a brief outline of how plate tectonics affected the distribution of continents and oceans through the Mesozoic.

The Mesozoic started with the Triassic period (252 to 201 million years ago), at which time all continents of the world were united into a giant C-shaped supercontinent called Pangaea, which had existed since about 300 million years ago (see figures 4 and 5). The northern limb of the C-shaped supercontinent consisted of the linked North American and Eurasian plates (together called Laurasia), while the southern limb comprised the joined plates of South America, Africa, Madagascar and India, Australia, and Antarctica (together called Gondwana). Between 250 and 200 million years ago, Laurasia and Gondwana were fully fused together in the west, where the North and South American plates touched with northwestern Africa and southwestern Europe.

On the inside of the C-shaped Pangaea supercontinent lay first the Paleo-Tethys Ocean, which closed over time (until 150 million years ago) and was replaced almost like-for-like by the Tethys Ocean, as we shall see below (see figure 4). The Paleo-Tethys and Tethys basins were shaped like a giant V lying on its side, roughly symmetrically about the equator, with the tip in the west and the opening in the east. During the first half of the Mesozoic, Paleo-Tethys was closing fast in a northward direction. It was subducted underneath Laurasia, as the Cimmerian plate—a long sliver of continent that separated Paleo-Tethys

to its north from the newly opening Tethys to its south—closed on
Laurasia by rotating from south to north across the equator like a giant door with a virtual hinge somewhere near present-day Spain (see
figure 4). The northward rotation of Cimmeria, and thus the demise
of Paleo-Tethys, resulted from rapid spreading of the Tethys basin to
the south of Cimmeria. Cimmeria fully collided with Laurasia at
around 150 million years ago, and a new subduction zone started to
the south of Cimmeria along the northern margin of Tethys.

Pangaea started to break up toward the end of the Triassic, from
about 200 million years ago. North and South America began to separate, and an inland ocean basin started to form between the African
and American plates, representing the very first stage of opening of
the North Atlantic basin. Via a passage between the North and South
American plates, this nascent ocean basin was connected with the
vast ocean on the outside of the C-shaped Pangaea supercontinent:
the Panthalassic Ocean. The rupturing was associated with massive
eruptions of the Central Atlantic Magmatic Province around 200 million years ago. These eruptions were a major driver behind the buildup
of the high-CO_2 greenhouse conditions of the Mesozoic. They coincided in time with a mass extinction that ended the Triassic period,
terminating half of all species on land and in the oceans.

During the Triassic, a noteworthy group of reptiles had developed
on land, from a lineage that had somehow managed to survive the
end-Permian extinction of about 252 million years ago. They were
called "synapsids" after their skull morphology, and they developed increasingly mammal-like characteristics. A branch of them gave rise to
the earliest mammals at around 225 million years ago (though some
suggest a more recent 166 million years ago). But the undisputed
rulers of the Triassic landscape were the archosaurs, which would
eventually give rise to crocodilians, birds, pterosaurs, and dinosaurs.
A severe reduction of non-dinosaurian archosaurs during the end-
Triassic mass extinction opened up the ecosystem for fast expansion of
the dinosaurs during the subsequent Jurassic period (201 to 145 million years ago).

Some studies have attributed the end-Triassic mass extinction (see
figure 1) to an asteroid impact. However, the prime suspect—a major impact crater in Canada—dates to around 214 million years ago,

and therefore is too old, while another crater in North Dakota seems to be of the right age, but is much too small for the type of impact that would have been needed. The more mainstream view relates the extinction to the epic volcanism of the Central Atlantic Magmatic Province. This is supported by the extinction's close association with a shift to low (more carbon-12) carbon isotope values. This concept attributes the extinction to what has been called a CO_2 supergreenhouse, and an acidification-caused marine biocalcification crisis. But other researchers emphasize that the isotope shift is too large to be explained by this mechanism alone, and favor a strong additional contribution from gas-hydrate breakdown. As with other extinction events, we understand the basic parameters, but not the full picture. The view is becoming increasingly prevalent, though, that the primary "kill mechanisms" of the end-Triassic and end-Permian extinction events may have been very similar, both related to major volcanic episodes.

Early mammals persisted through the Jurassic and subsequent Cretaceous period (145 to 66 million years ago), even though they remained small and hidden until the end-Cretaceous demise of the dinosaurs. Shortly after that event, placental mammals burst on to the scene in large numbers, diversifying into a wide range of sizes and shapes; the time after the Mesozoic would be the age of mammals, both of the older marsupial lineage and of the new placental lineage. Dominion of the oceans was handed back to fish; the mammalian whale and dolphin ancestor *Ambulocetus* only appeared as late as 50 million years ago.

You may wonder why allegedly "more evolved" mammals took so long, from 225 to 66 million years ago, before they replaced allegedly "less evolved" reptiles. But have you ever looked a large crocodile in the eye? Competition for ecological niches was somewhat unfairly loaded in favor of the reptiles during the Mesozoic. Remember, evolution does not care about working in a straight line along a preordained plan toward mammals and eventually humans, but instead proceeds via endless and seemingly haphazard experimentations, splits, and branchings, which all represent adjustments to fill available ecological niches with the best-adapted organisms of the time. Reptiles dominated virtually all ecosystems on the planet for almost 200 million years—on land, in water, and in the air. Many reptilian top predators even continued after the end-Cretaceous extinction, as in the case of crocodilians, snakes, and giant lizards (including Komodo dragons, and the five- to seven- meter-long *Megalania prisca* that lived in Australia until about 50,000 years ago). Reptiles appear to have been perfectly adapted to the world and opportunities of their time—arguably better so than mammals. They managed to sustain that advantage through-out the age of reptiles (252 to 66 million years ago), which lasted some three times longer than the entire age of mammals (66 million years ago until today). Our genus *Homo* only appeared on the scene as late as 2.7 or 3 million years ago, and our species *Homo sapiens sapiens* just a trifling 200,000 years ago, within which we have developed into a dominant species during the past 10,000 years or so (see figure 2). We've got quite a while to go yet before we could even consider claiming that we're more adapted or successful than giant reptiles.

The breakup of Pangaea that started 200 million years ago marked the beginning of the end of Tethys, before Tethys had even fully re-placed Paleo-Tethys by 150 million years ago (see figure 4). The fate of Tethys was sealed when subduction started along its northern margin after Cimmeria crunched into Laurasia. Its closure accelerated when India (initially together with Madagascar) separated from Africa, and these plates began northward journeys, from about 100 million years ago. Meanwhile, both the North and South Atlantic basins were open-ing fast. Most of Tethys was eventually crushed into oblivion as India collided with Eurasia from about 55 million years ago (though some say 35 million years ago). But a remnant can still be admired today. It is the eastern Mediterranean, which is going through its final stages of closure as Africa continues to rotate into Eurasia.

Throughout the extensive period of subduction of the low-latitude carbonate-laden sediments of the Paleo-Tethys and Tethys ocean ba-sins, there was strong volcanic CO_2 outgassing. This was another driver behind the Mesozoic high-CO_2 greenhouse conditions and their subsequent continuation until about 33.9 million years ago. But it is hard to say from the available evidence when exactly the Mesozoic CO_2 levels reached a maximum, as they were very high throughout. Values are often estimated at around 2000 ppm, and—less likely—up to 4000 ppm (see figure 11). Note, however, that a recent reassessment suggests that those estimates may be too high, and that values may have been closer to 1000 ppm. Although uncertainty remains, we can conclude that—on current evidence—Mesozoic CO_2 levels most likely sat somewhere in the range of 1000 to 2000 ppm, or two to three doublings higher than during our current interglacial before the industrial revolution (from 280 to 560, to 1120, to 2240 ppm).

In agreement with the generally high level of greenhouse gases, global temperatures were also high throughout the Mesozoic, albeit with considerable shorter-term fluctuations around that long trend. Note that we should not expect all temperature variability to always correlate perfectly with CO_2 changes. Additional variability is possible due to changes in Earth's energy balance related to changes in vegetation and land surface, in atmospheric dust content, in cloud cover, and in other greenhouse gases like CH_4, while flooding of shallow land areas also has an impact because water absorbs more heat than land surface. In any case, temperature reached a maximum in the middle of the Cretaceous period, roughly around 100 million years ago, when globally averaged temperatures were up to 10°C higher than the modern global average (see figure 14).

The Triassic had been a warm but in many places rather arid period, when high-latitude regions were covered in conifer forests with fern undergrowth. The subsequent Jurassic period saw global expansion of warm and humid conditions, which allowed lush broadleaf forests with ferns and cycads to spread into the high latitudes. The warm and humid conditions reached their apex at around 100 million years ago, in the middle of the Cretaceous period, after which a long and variable cooling trend set in that would ultimately lead to the current icehouse world. Overall, the Mesozoic tropics were warmer by several degrees, while the poles were much warmer than today—especially in the Middle Cretaceous. This means that there was a weak equator-to-pole temperature gradient, similar to what we saw earlier for the PETM. Deep-sea temperatures were some 8°C or 10°C higher than today.

Sea level had started off at levels similar to those of today during the Permian, and then kept on rising—with shorter-term fluctuations—until the Middle Cretaceous, when it reached more than 200 meters above the present. This is remarkable because only about 80 meters of sea-level rise might be explained if one were to melt all the ice on the world today and allow ocean water to warm up by 10°C. Clearly, anomalously high sea levels, high CO_2 levels, and high temperatures are among the defining characteristics to be explained for the Mesozoic. There is now little doubt that plate tectonics is at the heart of this explanation. In fact, the long-term developments through the Mesozoic era offer the most-mentioned example of how plate tectonics control greenhouse-gas levels and thus global temperature, as well as ocean spreading rates and thus global sea level.

Seafloor spreading rates were high during the Mesozoic, and most notably so during the Jurassic and Cretaceous, reaching maximum values at around 100 million years ago. There was an associated rise of sea level from near-present values during the end-Permian, through the Triassic and Jurassic periods, and into the Cretaceous period, until a Middle Cretaceous peak at around 100 million years ago. Notably, the onset of the Cretaceous period, 145 million years ago, was marked by particularly pronounced expansion of shallow seas and seaways. Between about 130 and 100 million years ago, much of Australia was covered by a shallow inland sea, called the Eromanga Sea, while North America contained a system comprising the Western Interior Seaway, Hudson Seaway, and Labrador Seaway. The Eurasian plate was similarly covered by enormous shallow continental seas, and the African plate contained large shallow seas as well. In all, about one-third of the modern land surface was submerged.

What all of the above amounts to, is—starting from about 250 million years ago—a period of about 150 million years with sustained high rates of ocean spreading, intense subduction of an ancient low-latitude ocean basin, major volcanism associated with development of new ocean spreading centers and subduction, and high sea levels that culminated at a few hundreds of meters above the present level in the Middle Cretaceous. This is a conspicuous combination that points at a single underlying mechanism: intense plate-tectonic activity.

Indeed, for the high Mesozoic seafloor spreading rates, we ought to expect sea level, subduction, CO_2 outgassing, and temperature to increase. This is because strong spreading is associated with hot and bulging spreading ridges, and therefore with increased ridge volumes.

At the same time, formation of new and additional spreading ridges also increases the spreading-ridge volume within the world ocean. All this ridge-volume increase displaces water in the oceans, which then can only go one way: up. To visualize this, fill a metal bathtub and mark the water surface with a line. Now take a sledgehammer and ram a major dent in the side of the bathtub to simulate the formation of new and bulging spreading ridges. Ensure that you make the dent without spilling any water. Then check the water level—it will have risen because the volume of the dent's inward distortion displaces water, which could only go in one direction.

Simultaneous with the trend to higher and higher sea levels, increased volcanicity associated with intense spreading and subduction causes intensified outgassing of external CO_2 into the atmosphere-ocean-biosphere system. Thus, the long trend of rising sea levels would be associated with a long trend of increasing CO_2 levels. And the latter, in turn, drives a long trend of warming because of greenhouse physics. In addition, long-term CO_2 outgassing would be associated with a long tendency toward acidification, were it not that the changes mostly were slow enough to be buffered by carbonate compensation. If and when we ever manage to reliably reconstruct CCD changes through these times, we may therefore expect to find a gradual upward shift.

Shorter-term variability should of course also be expected, superimposed upon the long-term trends. Any pulses of outgassing that were at times too fast to be directly offset by carbonate compensation would result in temporary, widespread ocean acidification spikes, with impacts on biota and with shorter-term fluctuations in the CCD—a bit like the 200,000-year fluctuation of the PETM. In addition, large and fast inputs of external CO_2 may drive oceanic low-oxygen crises along with the acidification spikes, as discussed before for the end-Permian and PETM events.

Interestingly, some researchers now argue that a low-oxygen crisis coincided with the end-Triassic event of acidification and supergreenhouse conditions, further supporting that the event was caused by massive volcanism. This then brings us to another extraordinary feature of the Mesozoic oceans; namely, an abundance of widespread low-oxygen or even anoxic (no-oxygen) events, called ocean anoxic events (OAEs). These events are among the most visually conspicuous (dark to black layers) and economically important (oil and gas) expressions of Mesozoic ocean variability. The following section will focus on OAEs, including similar events of smaller scale in the Mediterranean Sea, which happened much more recently and as a result have been more extensively drilled and studied.

CHOKING OCEANS

In a warmer world, like the Cretaceous and the Mesozoic in general, ocean density changes are less sensitive to changes in the surface freshwater budget (that is, the balance of freshwater in versus freshwater out) than in a colder world, like today. In addition, Mesozoic temperature contrasts over the globe were much weaker than today, and this is especially important when considering the equator-to-pole gradient. With much less cooling going on, surface-water density contrasts were much smaller across the Mesozoic oceans than today.

Density contrasts are needed to make new deep waters sink effectively. Deep-sea temperatures reflect surface temperatures in the regions where the new deep waters are sinking, because temperature inside the ocean changes only with mixing. Therefore, deep-sea temperatures of 10°C to 12°C indicate that surface waters in the Mesozoic regions of new deep-water formation did not cool much below these levels. This immediately indicates that there was no sea-ice formation to help drive up the water density. The prevailing moderate temperature contrasts would drive only moderate density increases, relative to what happens in a cold world like today. In addition, any salinity increase would have been less effective at driving up density in a warm world, relative to the impact of a similar salinity change in a cold world. In consequence, we might expect deep-water circulation to be weaker in a warm world compared with a cold world, although recent computer modeling suggests that this was not necessarily the case.

Oxygen is more soluble in water at low temperatures than at high temperatures; from 0°C to 30°C, the oxygen solubility drops by almost 50%. This means that, for a given atmospheric oxygen concentration, cold water can hold more oxygen than warm water. When Mesozoic deep waters were formed at 10°C to 12°C, they started out with 25% to 30% less oxygen than modern deep waters that form at about 2°C. In consequence, deep-sea oxygenation in a warm world may have taken a double hit from both weakened deep-water formation and circulation, and lower initial oxygen concentrations within the new deep waters. And then the hit may have been triple: for every 10°C warming, the rate of life processes (metabolism) doubles, so that warmer deep waters would have caused increased oxygen utilization during respiration/decomposition. To see if this triple hit indeed took place, we first need to specify our expectations about Mesozoic ocean circulation and oxygenation in more detail, so that we can then compare these with the available evidence.

During the greenhouse times of the Mesozoic, culminating through the Jurassic in the Middle Cretaceous, the equator-to-pole sea-surface temperature gradient appears to have become less than 10°C, compared with today's gradient of 30°C. This reduced gradient mainly resulted from much warmer poles. Middle Cretaceous high-latitude sea-surface temperatures as high as 25°C to 30°C have been reconstructed, but these may represent the growing season at those latitudes for the algae and zooplankton whose remains are used to determine the temperatures. That growing season likely was the summer half year, given that high latitudes receive little to no sunlight for primary production during the winter half year. Also, there remains strong debate about some of the evidence used to infer the very high-end temperatures—notably about those based on the temperature-sensitive chemical structure of archaean cell membranes. Winter temperatures may have been considerably lower, as is indicated also by the lower deep-water temperatures reported for the time.

Deep-water temperatures of 15°C or even 20°C have been suggested for the Middle Cretaceous. This means that—at least in the cold season—such temperatures must have existed in surface waters at the sites of new deep-water formation, given that this is where deep-water properties are set. A picture emerges in which surface temperatures based on algal and zooplankton data may have reached as high as 25°C to 30°C at high latitudes in summer, while winter temperatures were some 10°C lower than that. Therefore, it appears that there was enough seasonal cooling to generate the density increases needed to effectively drive new deep-water formation. After all, this inferred surface-temperature variability is more pronounced than that in today's Mediterranean and Red Sea, where a good supply of new deep water is produced.

The next issue to be dealt with is salinity. The higher the salinity, the more impact a given temperature change would have on density, although this salinity influence vanishes at temperatures above 12°C. To make a salinity estimate, a couple of things need to be considered.

First, major salt depositions have taken place in the Mediterranean, Red Sea, and Persian Gulf regions, between 13 and 5 million years ago, as we shall discuss later. The original deposited mass of sea salt in these joint deposits was about 3.66×10^{18} kilograms, or 3.66 million billion metric tons. The mass of global ocean water is about 1.35×10^{21} kilograms. The removal of salt into these deposits therefore represents a world-ocean salinity reduction of about 2.7. Remember, salinity is reported in numbers without units, as salinity is measured by relative conductivity, but in a practical sense this number represents about 2.7 grams of salt per kilogram of seawater.

A second influence arises from the fact that there was less continental ice on the planet than today. In the Middle Cretaceous, there was no significant land ice at all. Addition of the entire modern volume of land ice to the world ocean would drop salinity by about 0.7. Around 10 million years ago, the volume of ice was roughly half what it is today. For 10 million years ago, therefore, we can estimate average ocean salinity as 2.7 − (0.7/2) = 2.4 higher than today. Today, average ocean salinity is 34.7, so our calculation suggests that it was close to 37 at 10 million years ago. By extension, we might then think that salinity about 100 million years ago differed from that of 10 million years ago only by addition of the other half of the ice volume, in which case salinity would have been about 36.7. Unfortunately, things are not that simple when looking over such a long expanse of time. This is because erosion and weathering of older salt deposits and other rocks slowly but surely add salts to the ocean, which creates a slow and steady increasing trend in salinity over time. It is estimated that this caused a salinity increase of about 3 over the time span between 100 and 10 million years ago. Thus, we estimate average ocean salinity for the Middle Cretaceous at roughly 36.7 − 3 = 33.7. Let's say 34. The history of past ocean salinity has been studied for many decades, and a comprehensive assessment was made as recently as 2006 by the American geoscientist William Hay and colleagues. The numbers I have used above represent a simplified extract from their work.

At an inferred salinity of about 34, Middle Cretaceous ocean waters were remarkably similar to modern ocean waters. Hence, the sensitivity of dense-water production to freezing, evaporation, and cooling was rather similar to what we see today too. We can look at the modern Mediterranean and Red Sea to see how deep-water formation can be driven in the absence of freezing conditions. In those basins, cooling of the warm ocean surface is strongly dominated by latent-heat loss during evaporation. Evaporation not only cools the water, but also increases its salinity—this gives a double impact on density. Consequently, evaporation and its coupled cooling likely were key processes for new deep-water formation in the Mesozoic, much like they are in the modern Mediterranean and Red Sea. To summarize, the Mesozoic oceans were most likely filled with—for those times— relatively cool deep waters with salinities that were roughly average or just above average, which most likely originated from high latitudes. Many recent studies concur that a scenario with deep-water formation in high latitudes of the North Pacific and Southern Ocean is more likely than the previously long-held concept of low-latitude formation of salty and warm deep waters.

There will have been considerable consequences for deep-water oxygenation. Gas exchange is primarily affected by temperature, and to a lesser extent by salinity. If Middle Cretaceous deep waters were formed at 15°C to 20°C, then their oxygen concentrations at source would have been up to 30% lower than in modern newly formed deep waters, just because of the temperature difference alone. For each five parts per thousand of salinity change, oxygen concentrations at equilibrium with the atmosphere change by only 3%. Given that newly formed Middle Cretaceous deep waters would have had roughly similar salinities to modern newly formed deep waters, the salinity influence on oxygen concentrations was tiny. In consequence, we find that Middle Cretaceous waters were disadvantaged for oxygen by up to 30%, relative to their modern counterparts. That implies a substantial drop in oxygen supply to the deep sea. Now let's compare these theoretical expectations with the available observations.

Several times during the Mesozoic, large parts of the ocean became anoxic, which means that the waters no longer contained oxygen. These OAEs have actually happened several times during Earth's history, not just during the Mesozoic. But the Mesozoic examples—especially those from the Cretaceous—are well preserved in ocean sediments and have grown into great economical prominence because they have often developed into oil and gas source rocks. These attributes have driven an exceptional research interest in OAEs.

Seafloor sediments deposited during OAEs stand out by virtue of their very dark color and rich organic-carbon content, which commonly reaches between 1% and 10% by weight. For comparison, regular marine sediments typically contain only 0% to 1%. OAE deposits, also called black shales, are the source rocks for the world's main crude oil reserves. During the Mesozoic there were three particularly notable anoxic events, in addition to a number of more minor ones. One of the major events took place at around 183 million years ago, in the Jurassic. The other two major events occurred during the Middle Cretaceous, and represent the best-studied examples that we focus on here.

One of the two major Cretaceous OAEs considered here happened at around 120 million years ago, in the early Aptian age. The other occurred at around 93 million years ago, at the boundary of the Cenomanian and Turonian ages. The Aptian event is known as OAE-1a. The Cenomanian-Turonian event is known as OAE-2.

Geographically, the Mesozoic OAEs are best known from the tropical and subtropical Tethys Ocean, and from the North and South Atlantic Oceans that were beginning to open at the time. It is strongly debated whether OAEs were truly global or not. To understand why this debate rages on, we need to consider the controls on deposition and preservation of large quantities of organic matter.

Under oxygenated conditions, organic matter is effectively decomposed during its fall through the water column, and also on the seafloor for particles that have reached the bottom, until they are sufficiently deeply buried in the sediments. Two main processes can overwhelm the oxygen supply in deep waters and thus lead to the preservation of organic matter at the seafloor, which increases the quantity buried in sediments.

One process is a loss of oxygen in the water column. An important cause of this is a change in environmental conditions in the areas of new deep-water formation that reduces or stops deep-water formation. In the absence of new deep-water formation and circulation, the water column becomes stratified, and the deep waters are no longer renewed, and thus no longer supplied with oxygen. We call a deep-water mass that has stopped circulating a "stagnant" water mass.

Along with many people, I have a bit of personal experience with a small-scale example of this; namely, with my fishpond, which warmed up in a hot summer, so that its deeper waters were no longer renewed and became stagnant. When that occurs, ongoing decomposition of organic matter—which, to make matters worse, is accelerated when rates of life processes (metabolism) go up with rising temperature—then rapidly depletes the oxygen. Once this happens to your pond, your fish will come to the surface, where there is still a bit of oxygen, then they start gulping in water and air to try to get enough oxygen, and eventually they die (I certainly did not receive a note of endorsement from my fish that summer). In the worst cases, the pond may be taken over by cyanobacteria, which can thrive in anoxic conditions, and conditions in the deeper waters can become euxinic, which means that there is free hydrogen sulfide (H_2S) in the water column. You'll know that has happened when you stir the water a bit and you get a whiff of the unmistakable rotten-eggs smell of H_2S. H_2S is highly toxic to most organisms, except for a few extreme specialist microbes. In some OAEs, organic molecules have been found that come from such extreme specialists, which are adapted to live only in euxinic conditions. This tells us that conditions at times became even worse than anoxic, which is a pretty amazing sign of deterioration when considered on a scale as vast as an entire ocean basin. But we are pretty sure that it happened, and not just once. Once anoxic conditions take hold in a water body, things can quickly go completely off the rails.

The other main process that helps to deplete oxygen from the water column is the rate of organic production. When that increases, typically because of an increase in nutrient supply, then the rate of downward transport of organic matter increases. In this case, oxygen is depleted faster in deep water because of an increased oxygen demand for decomposition. New deep-water formation does not have to stop for this process to deplete all the oxygen; the only thing needed is that the rate of oxygen utilization in decomposition overwhelms the rate of new oxygen supply by new deep-water formation. Here, the doubling in oxygen-demanding rates of life processes in animals for every 10°C of warming becomes an important accelerating influence. In the rather warm Mesozoic deep sea—with all else remaining exactly the same—even that influence alone would have made oxygen demand almost twice as high as it is in the modern, cold deep sea.

Most people are familiar with an everyday example of factors that overwhelm oxygen supply; it has been in the news a lot. Today, fertilizing of agricultural fields and drainage of phosphate-based detergents cause increasing nutrient concentrations in streams, rivers, lakes, and coastal seas. The oversupply of nutrients into those systems, called eutrophication, causes increased primary production, and thus an increase in oxygen demand through decomposition. Even in actively circulating systems with ongoing oxygenation, oxygen demand can come to outweigh oxygen supply, resulting in low-oxygen to anoxic or even euxinic conditions. One of the more dramatic expressions of human-induced eutrophication is the massive expansion since the 1970s of so-called dead zones in coastal regions and large lakes. These zones form when eutrophication causes massive blooms of algae and harmful cyanobacteria, which trigger survival crises for other organisms because of the large-scale oxygen depletion caused by decomposition of the sinking organic matter. The Great Lakes in North America have been getting a lot of press in recent years because of toxic algal (cyanobacterial) blooms, but the problem is much more widespread around other freshwater systems and global coastlines.

The two main processes that lead to oxygen depletion commonly operate closely together. One restriction to that collaboration, however, is that productivity cannot continue or increase if deep-water circulation is stopped completely, as nutrients released into deep water by decomposition need some circulation to be transported back up into the photic layer. In the absence of deep-water circulation, only external nutrient supply can keep production going, which is possible near rivers or other sources of new nutrients, but not on an ocean-wide scale. In other words, to develop basin-wide anoxia with burial of significant amounts of organic matter, some vestiges of deep-water circulation remain necessary to redistribute nutrients. The oxygenation that comes associated with that circulation will be most notable close to the areas of new deep-water formation, and then will rapidly reduce away from those areas because of oxygen demand for decomposition. This means that regions of new deep-water formation are unlikely to have become fully anoxic, and this reasoning underlies the debate about whether OAEs were truly global or not.

One further factor helps a lot in the development of anoxia in the oceans; namely, geographic near isolation of a basin from the open ocean. A restricted basin has only limited exchange of water masses with the open ocean, and thus—often—rather poor oxygen supply into its deeper waters. In addition, runoff from landmasses around a restricted ocean basin can supply nutrients, while existence of some deep-water circulation and wind-driven upwelling of subsurface waters can redistribute those nutrients through the basin for maximum effect on production. Furthermore, the proportionally large expanses of shallow margins around such basins imply short pathways for organic matter sinking to the seafloor, while runoff from the surrounding continents supplies a decent amount of sediment for burying the organic matter. Together, this means that more organic matter hits the seafloor than in the very deep ocean, and also that it gets buried more efficiently in the sediments than in the deep open ocean. In the Mesozoic, the newly opening North and South Atlantic Oceans were restricted basins in this sense, and as such they were perfectly conditioned for organic-matter burial; they were prime sites for Mesozoic formation of black shales.

We are fairly confident that Cretaceous OAE-1a and OAE-2 played out more or less as follows.

There was already strong outgassing during a period of high plate-tectonic activity as the underlying process behind the development of the broad Middle Cretaceous thermal maximum, but osmium isotope research indicates that volcanic activity started to intensify even more from about 300,000 years prior to OAE-1a. Then, about 180,000 years before OAE-1a, exceptional amounts of volcanic CO_2 were emitted during the formation of the Ontong Java Plateau (see figure 13), which drove both a widespread multistage shift toward low carbon isotope values, and a strong increase in CO_2 concentrations. The latter, in turn, drove sharp warming by 1.5°C to 2°C, which peaked several tens of thousands of years after the onset of OAE-1a. Some studies report a potential ocean acidification impact at this time of major volcanic outgassing.

Meanwhile, fossil assemblages indicate that continental runoff, together with increased input of metals from the volcanic activity, increased nutrient supply to the oceans and that primary production increased when the volcanic event and warming were most intense. OAE-1a lasted 0.5 to 1 million years. With some fluctuations, a cooling occurred through the time of the OAE-1a event under enhanced organic production. This suggests that net carbon burial (formation of the black shale) related to the high organic production extracted enough carbon from the atmosphere-ocean-biosphere system to drive a cooling trend. This scenario is supported by association of the cooling with a gradual return to higher carbon isotope values owing to burial of carbon-12 enriched marine organic matter. High production seems to have been maintained despite extensive organic matter (and thus nutrient) burial, through intensified external-nutrient input due to intense weathering under warm, high-CO_2 conditions, aided by the exposure of easily weathered, fresh volcanic rock.

OAE-2 also followed an episode of increased volcanic activity, as indicated by a diverse array of information, including trace-metal, osmium isotope, and strontium isotope data. In particular, there was the formation of the large Caribbean plateau and Madagascar flood basalts, which started less than a million years before the onset of OAE-2. Once it had been initiated, OAE-2 lasted between about 250,000 and 700,000 years. By comparing the onset of organic-matter burial with the timing of the carbon isotope shift, it has been inferred that low-oxygen to anoxic conditions existed even before OAE-2 in the southwestern North Atlantic, and that these intensified and expanded throughout that basin as well as the southwest Tethys Sea, and also into the South Atlantic and early Indian Ocean at the onset of OAE-2. Presence of organic-rich sediments suggests that (parts of) the equatorial Pacific were affected, and model-based work even proposes that large expanses of the entire Pacific may have been involved. These same model-based assessments suggest that these regions were oxygenated prior to the onset of OAE-2, and that the overall global oceanic oxygen content became reduced by more than 70% during OAE-2, with true anoxia expanding from 5% to more than 50% of the ocean volume. Notably, both observational and modeling studies infer that the deep ocean experienced large variations in oxygenation, so that not all places were continuously anoxic.

Atmospheric CO_2 levels peaked just before OAE-2 and during its early phase, reaching values estimated at 500 to 3000 ppm (generally, studies seem to favor values in a range of 1000 to 1500 ppm—it's quite uncertain). At the same time, subtropical and tropical sea-surface temperatures rose by 5°C or more, to values in the region of 33°C, or even 42°C in more extreme results. Values in the high northern latitudes may have reached 20°C or so. As with OAE-1a, the onset of carbon burial into sediments was associated with the onset of a drop in CO_2 levels.

The existence of anoxic conditions and even black-shale deposition in the southwestern North Atlantic before OAE-2 has been attributed to the restricted nature of basins in that region. Seafloor ridges reduced the connectivity with deep-water circulation in the rest of the (open) ocean, and there were no local sites of new deep-water formation.

Modeling experiments suggest that oceanic concentrations of the key nutrient phosphate before OAE-2 were similar to those of today. They likely doubled during the event, causing a proportional increase in productivity. Increased weathering due to warming may have provided the initial nutrient increase, and phosphate release (referred to as nutrient regeneration) from anoxic sediments—brought back toward the sea surface through circulation—would have maintained high levels throughout OAE-2. Nitrogen likely was obtained from nitrogen fixation using dissolved gas in surface water (which equilibrates with the air), a trick mastered by *Trichodesmium*, a genus of filamentous cyanobacteria. A similar change in nutrient cycling is thought to have characterized other OAEs and also smaller anoxic systems, including the much younger anoxic events that happened in the Mediterranean.

There is much more information about the Mesozoic OAEs than that summarized above. However, it remains incredibly difficult to place what happened in the full context of a basin with adjacent land-masses, and with clearly defined processes that link the black-shale deposition to circulation, oxygenation, and productivity changes, and eventually to climatic variations. To obtain that sort of information about potential processes and controls, we may consider the more recent anoxic events in the Mediterranean, which developed during the most recent 15 million years. So, even though it breaks us tem-porarily away from the main chronological flow of the history of the oceans, the remainder of this section looks at what these Mediterra-nean miniature analogues can tell us about mechanisms that may help explain older OAEs.

*****

In the Mediterranean Sea, and especially in the eastern Mediterra-nean (to the east of the Strait of Sicily), anoxic events have taken place with great regularity during the past 15 million years or so. These were events in a small oceanic basin, rather than on a near-global scale. They have been recovered in hundreds of sediment cores through-out the basin, and can also be studied in marine sediments that have been uplifted and are accessible on land in many countries around the basin. Moreover, the most recent Mediterranean anoxic event hap-pened as recently as only 10,000 to 6,000 years ago, which is young enough to allow dating at high precision using radiocarbon.

The Mediterranean anoxic events resulted in deposition of dark sediment layers that contain more organic carbon than the surround-ing sediments. These organic-rich layers were named "sapropel," from the Greek words for putrefaction and mud, but might as well have been called black shale. All around the Mediterranean Sea, sapropels exist within otherwise white, light gray, and yellow marl sequences. Yet it was not until the Swedish Deep-Sea Expedition of 1947–49 that sediment cores taken from a research vessel revealed the basin-wide nature of these sediments within seafloor sediments.

Those cores and many others from subsequent expeditions established that the most recent sapropel of about 10,000 to 6000 years ago was just one example of a regular sequence. The more recent sapropels were intensively studied from shallow-seafloor cores taken with many nations' regular research vessels, and older examples were studied intensively from sediment sequences now uplifted on land. What was not yet known was how the older examples in land sections related to the younger examples found in sediment cores from throughout the basin. Deep-seafloor drilling by the Deep-Sea Drilling Project in 1970 and 1975 established that the sapropels exposed on land indeed continued through the basin in the seafloor sediments. This also demonstrated that the remarkably regular spacing (in time) of sapropels spanned millions of years.

Most Mediterranean sapropels were formed recently enough that the basin in which they formed looked more or less the same as it does today. This removes uncertainties about exact depths, presence of underwater barriers, and so on, which always plague research of events in the deeper past. Thus, studies of Mediterranean anoxic events can provide a level of spatial and time-dependent detail that is impossible to achieve for Mesozoic OAEs. In the 1980s, the French geoscientist Martine Rossignol-Strick established that sapropels formed regularly because they were triggered by maxima in northern-hemisphere summer insolation, as timed by the orbital cycles of precession and eccentricity. The Dutch geologists Jan-Willem Zachariasse and Frits Hilgen led a decades-long international effort to meticulously piece together many land-exposed sections into a continuous time series. Hilgen then used the relationship between sapropels and astronomical variability to develop a highly detailed "astronomically tuned" time scale for the past five million years, together with Lucas Lourens and the essential astronomical solutions calculated by the French astronomer Jacques Laskar. Renewed deep drilling by the Ocean Drilling Program (ODP) in 1995 provided much additional detail to this effort. The astronomically tuned time scale has since been extended further back in time, both in the Mediterranean and outside it. In addition, recovery of the ODP (now the International Ocean Discovery Program, IODP) cores, along with many shorter cores in high spatial density by regular research vessels, has allowed the establishment and testing of hypotheses about the controls on sapropel deposition.

Sapropels are found from about 300 or 400 meters' water depth down to the greatest depths in the open basin, which reach more than 5000 meters. In restricted marginal seas like the Aegean Sea between Greece and Turkey, sapropels exist anywhere below about 120 meters' depth. Through time, sapropels developed more frequently in the eastern Mediterranean than in the western Mediterranean—these two basins are separated by a ridge, or sill, of only 440 meters' depth in the sea strait between Sicily and Africa. The organic-carbon content of sapropels typically ranges between 2% and 8% by weight, but some of the more recent examples go up to 14% or 15%, while some sapropels that formed between two and three million years ago have a ridiculously high organic-carbon content of about 30%. In geologic terms, sapropels are quite young, and thus have not had much time to be "matured" into petroleum or gas. But new finds in the eastern Mediterranean indicate that much more gas has been formed from them than had been anticipated. As a result, they are fast gaining in importance to the fossil-fuel industry.

Sapropels developed in response to reduced oxygen supply by reduced deep-water circulation, increased oxygen demand due to organic productivity increases, or some combination of these. The relative importance of these controlling processes appears to have varied between sapropels. Another important aspect is the geographic setting, and especially the degree of basin restriction. This aspect in particular helps to explain why sapropels formed more frequently in the eastern Mediterranean than in the western basin, or formed to shallower depths in restricted marginal seas than in the open basins. All these controls are pretty similar to what we discussed for black-shale deposition during OAEs. However, there are clear differences too. The main one lies in the fact that sensitivities to the various controlling processes were different for sapropels than for OAEs because the basins in which they formed were differently shaped, had different volumes, and were differently affected by environmental changes. Also, Mediterranean sapropels formed over periods of a few thousand years, while OAEs lasted hundreds of thousands of years. And finally, the quantity of carbon extraction during OAE black-shale deposition was sufficiently large and lasted sufficiently long to affect global climate. This was not the case for Mediterranean sapropels.

As we have seen before, sapropels formed during maxima in northern-hemisphere summer insolation. At these times, the African monsoon was intensified to such an extent that the Sahara Desert of North Africa was greatly reduced by northward expansion of the savanna and woodlands. But let's begin at the beginning: what is a summer-insolation maximum and why do they occur every 21,000 years or so?

The orbital cycle of precession causes climate changes in a cycle with two main periods of 19,000 and 23,000 years; for convenience, we simplify this to a cycle with an average period of 21,000 years. Every 21,000 years we go through a change from a precession maximum (the modern configuration) to a precession minimum, and back again. This means that half a precession cycle ago, roughly 10,000 to 11,000 years ago, Earth was at a precession minimum. During a precession maximum, northern-hemisphere summer falls close to the point of aphelion in the Earth's elliptical orbit around the Sun; that is, the point when Earth is furthest from the Sun. During a precession minimum, northern-hemisphere summer falls close to the point of perihelion in the Earth's orbit; that is, the point when Earth is closest to the Sun. The situation with respect to aphelion and perihelion is exactly opposite for northern-hemisphere winter. In consequence, seasonal contrasts are enhanced on the northern hemisphere during a precession minimum, and weakened during a precession maximum.

When Milanković calculated the effects of the orbital cycles on insolation in different bands of latitude, he found that precession minima were marked by a northern-hemisphere summer-insolation maximum and winter-insolation minimum. This is mainly because of precession, and the intensity of each precession minimum (and thus insolation maximum) is modulated by the degree of eccentricity of the Earth's orbit. The more elliptical the orbit, the greater the impact of precession. During a high-amplitude precession minimum—that is a precession minimum during a time with a most elliptical orbit—the northern hemisphere experiences a more intense insolation maximum than in a precession minimum during a time with a near-circular orbit. The orbital eccentricity changes over time scales of 100,000 and 400,000 years.

Astronomers, in particular Jacques Laskar, have focused on determining the essential precise astronomical solutions as far back in time as possible. Unassisted, the solutions have a limit of prediction at about 60 million years, because of chaotic elements in the calculations that are related especially to the orbits of the dwarf planets (giant asteroids) Ceres and Vesta. Assisted by constraints to the solutions from geologic data, however, the solutions could be extended. For detail on the scale of the precession cycle, the method realistically is accurate only back to five or six million years ago. But viewed at the scale of the 100,000-year eccentricity cycle, it currently extends to 50 million years ago, and use of the stable 400,000-year eccentricity cycle might allow it to be used to 250 million years ago or so.

The sapropel pattern of the past 15 million years clearly follows insolation variations. They are fundamentally spaced every 21,000 years in relation to precession, but are often missing during times when the summer-insolation maximum was not very intense because of a minimum in eccentricity (near-circular orbit). In times of maximum eccentricity, sapropels are well developed. Consequently, we find clusters of well-developed sapropels spaced 21,000 years apart during eccentricity maxima, and only poorly developed sapropels or absence of sapropels during eccentricity minima. Whenever a summer-insolation maximum is strong enough, a sapropel typically starts close to the time that the insolation maximum is reached. Such regular "clockwork" behavior requires explanation.

It is the African monsoon that links insolation to deep-water oxygenation changes in the Mediterranean. The critical issue is the increasing seasonal contrast on the northern hemisphere during precession minima. Ocean water has a high thermal capacity, which means that it follows seasonal temperature changes slowly, with relatively small amplitude, and with a delay of one to two months. In contrast, land surface has a low thermal capacity; it follows seasonal changes rapidly, with large amplitude, and with hardly any delay. We all know this from experience—seasonal temperature variability is relatively subdued in regions with a maritime climate, like northwestern Europe, but is large, between strong summer and winter extremes, in regions with a continental climate, like interior Canada or Siberia. As a result of the different thermal capacities, land-to-sea contrasts in temperature increase during times with strong seasonal contrasts, notably during precession minima.

Monsoons are fundamentally driven by land-sea temperature contrasts. When land heats up quicker and more than the sea, low pressure develops over land, and higher pressure over the sea. Surface airflow is then established from high-pressure to low-pressure regions, and over the sea this airflow becomes laden with moisture from evaporation in summer. As the air is transported onto land, and is then lifted up in the low-pressure region, it expands and therefore cools, the moisture condenses out, and rain develops.

As a consequence of this fundamental underlying drive, the African monsoon was intensified during precession minima, along with all northern-hemisphere monsoon systems. Intensification of the African monsoon circulation led to more rainfall over, and river runoff from, areas that are also affected by the African monsoon today. Notably this concerns the Ethiopian highlands, from which monsoon rainfall drains via the Blue Nile and Atbara rivers into the main Nile, and through that into the Mediterranean.

Besides gaining in intensity, the monsoon also pushed much further northward over Africa. This led to something that seems almost unimaginable today: greening of much of the Sahara Desert region. Savanna and woodlands from the south pushed northward by 1000 kilometers during weak "green-Sahara" events, and up to 2000 kilometers during strong ones. This strongly compressed the arid desert itself into a narrow band located close to the modern Sahara's northern margin, or even erased it entirely. The monsoon rains fueled major lakes and (likely seasonal) river systems in places that today are among the very driest on Earth, such as the Oyo depression in Sudan. Lake Chad grew to many times its current size; we refer to it as "Megachad."

During strong green-Sahara episodes, monsoon rains extended to the north of the central Saharan mountains, which form the watershed between drainage to Lake Chad in the south and to the Mediterranean in the north. As a result, major rivers, likely seasonal, flowed northward into the Mediterranean Sea in various places along the North African margin. Some of these rivers have been imaged with ground-penetrating radar from the Space Shuttle. In addition, people such as Hans-Joachim Pachur from Germany, Nick Drake from England, Françoise Gasse from France, and many others have spent decades of fieldwork physically mapping the Sahara's streambeds and lakes. This work is not straightforward, as conditions are harsh, political tensions abound, and many of these studied systems are currently—during a precession maximum—well hidden under the desert's shifting sands.

The green Sahara's massive new savanna and woodland area during intense precession minima became a haven for wildlife, including prehistoric humans. People have left rock art. Animals have left their bones, from which we know that buffalo, giraffes, hippos, and crocodiles were present in the green Sahara in great numbers. This list includes species that require year-round water in lakes or rivers. It has also been suggested that the regularly recurring cycle of expansion and contraction of green-Sahara conditions repeatedly created attractive opportunities for expansion, followed by crises that called for adaptation, and thus may have played an important role in shaping modern humans' adaptability and evolution.

In sediment cores from oceans around North Africa, green-Sahara periods are recognized by massively reduced input of windblown dust, and in the Mediterranean also from river-borne oxygen and neodymium isotope signals. Thus, we are confident that sapropel deposition in the Mediterranean coincided with green-Sahara events. The two are linked by enhanced monsoon runoff that was channeled into the Mediterranean through the Nile and—now dry—rivers along the wider North African margin. This large inflow of fresh water in turn wreaked havoc with the Mediterranean deep-water circulation.

The subtropical Mediterranean Sea has its own sources of new deep-water formation. Water flows into the basin at the surface through the Strait of Gibraltar, and strong evaporation from the basin creates high salinity by the time the water makes it to the eastern margin. Along the way, the water also warms up. Between the islands of Rhodes and Cyprus, winter cooling more or less removes the excess temperature that has built up in the surface waters, and this causes formation of a new intermediate water mass that is relatively warm at 15°C to 16°C, and very saline at almost 39. The temperature of intermediate water is similar to that of the water flowing in through the Strait of Gibraltar, but its salinity is three parts per thousand higher; hence, it is evident that its formation is primarily driven by the salinity increase.

Mixing of surface and intermediate water with some additional cooling and a small reduction in salinity causes new deep-water formation in the Aegean Sea and Adriatic Sea, and also in the northwestern Mediterranean to the south of France. The deep waters in both basins are cool, with temperatures between 10°C and 13°C, and also retain a relatively high salinity of roughly 38.5. Both in the western and in the eastern Mediterranean, deep-water circulation is strong today, and deep waters are well oxygenated as a result.

During times of intense monsoon flooding from the Nile and wider African margin, the salt gain that drives the first stage of deep circulation—that is, intermediate-water formation—is disturbed. This is because evaporation creates a salt gain, but river input reduces it again. Today, the intermediate water supplies salt to the regions where deep waters are formed, and salt is of prime importance to that process. Without the salt, deep-water formation becomes disturbed because cooling alone is not enough to generate new deep water. When this happens, old deep waters are no longer replaced. Oxygen supply to the deep sea then stops, and even unchanged levels of organic production will create enough oxygen demand over time to turn the deep waters anoxic.

To make matters worse, productivity also increases, as documented by a host of fossil evidence for changes in the operation of the ecosystem. Initially, this increase is driven by nutrient input by the rivers. Later it is boosted and maintained by phosphate regeneration from sediments and nitrogen fixation from dissolved gas in surface water (which had equilibrated with the air), just as we saw in the case of the black shales. Often, such concepts were first invoked, developed, and tested with numerous types of measurements in the easily accessible Mediterranean sapropels. In any case, increased productivity causes increased oxygen demand in the deep waters. Eventually, the Mediterranean deep sea becomes anoxic—severely so below about 1800 meters' water depth, and less intensely so at shallower levels. The anoxic episodes typically last a few thousand years.

In some sapropels, occasional cooling events over a few centuries up to about a thousand years have caused temporary reoxygenation events down to about 1800 to 2500 meters' depth. Similar reventilation events have been recognized in OAE black shales, although they may have lasted longer in those cases. Reventilations are a strong indication that some new deep-water formation and consequent deep-water circulation persisted through the anoxic events. As discussed before, this is a prerequisite for the nutrient regeneration that is needed to keep the events going.

Some sapropels contain organic molecules that indicate expansion of euxinic conditions all the way into the photic layer. This is a very serious environmental deterioration, and it is only found in intensely developed sapropels. The same organic molecules are encountered at times in OAE black shales, which again indicates a strong parallel between sapropels and black shales. In consequence, sapropel and black-shale studies keep developing in parallel, with lots of cross-fertilization of concepts, hypotheses, and methodologies. The key differences, as mentioned before, are differences in scale and duration, and of course the areas in which OAEs formed were much larger and less limited to a single semi-isolated basin.

The above gives a reasonable overview of the controls on OAEs. The discussed increases in the preservation of organic matter through low-oxygen conditions, and in the rate of organic transport to the seafloor through increased productivity, together lead to increased organic-carbon burial in marine sediments. Thus, black shales or sapropels are formed, and subsequently become the source rocks for many commercial oil and gas deposits. Once the petroleum products are cooked out of the source rocks by geothermal heat, deep in the sediment bed, oil and gas typically collect into more porous sediment layers, which we call reservoir rocks. But the products would ultimately escape from those reservoirs if there were no impermeable seal to lock them in: the cap rock. This seal often consists of a layer of impermeable clay or—especially for some of the highest-yielding fields—impermeable salt.

The presence of salt in marine sediments implies that seawater salinity must have risen tenfold or more, which is nearly unfathomable in an open oceanic setting. Regardless, massive salt deposits have formed quite a few times during Earth's history. We refer to such ocean-basin-scale salt deposits as salt giants. The next section will focus on salt-giant formation within the Mesozoic, in both the newly forming North and South Atlantic basins, and on a much younger event of six million years ago in the Mediterranean Sea. As with the OAEs, we will first discuss key features of salt giants in general and of Mesozoic events in particular, and will then use the much more recent Mediterranean event to obtain deeper insight into the likely processes involved.

SALTY GIANTS

Sometimes in Earth's history, exceptional things have happened that surpass our wildest imaginations. For me, the deposition of salt giants, also known as saline giants, is right up there among the most unimaginable of events. These deposits are gigantic masses of evaporites—the deposits that are left when salty water is evaporated away. What's so special about the salt giants is that they are enormous. They span areas of hundreds of thousands to more than a million square kilometers, and are several hundred to a few thousand meters thick. They contain gargantuan amounts of salts.

Evaporation of seawater produces a sequence of evaporites. Typically, we first see calcite (calcium carbonate) and dolomite (calcium-magnesium carbonate), followed by the true evaporites. The latter go from first deposition of gypsum and anhydrite (calcium sulfate, $CaSO_4$), via the common rock salt or halite (sodium chloride, NaCl), to a range of more exotic potassium, magnesium, and sodium sulfates and chlorides. These more exotic salts are often referred to as the bittern salts, after the fluid that remains in salt pans when halite has been harvested. From evaporation of a 1000-meter column of common seawater, we would get 17 meters of evaporites; namely, 0.6 meters of gypsum, 13.3 meters of halite, and 2.7 meters of bittern salts.

So here's a bit of a conundrum: how can evaporite sequences be hundreds to a few thousand meters thick if evaporation of even three kilometers of seawater would produce only about 50 meters' worth of salts, and the basins in which the evaporites were formed were often not that deep to begin with? A cut-off basin could indeed dry out, but then there would be just a single sequence of evaporites, with a thickness of at most 1.7% of the depth of the basin that dried out. It can only be that the basins in which the evaporites formed were not completely cut off from the open ocean, so that new supply of ocean water occurred to feed the evaporation, and to thus provide the extraordinary thicknesses of deposits that we observe.

To create a sequence of evaporites that is 1000 meters thick, a basin 1500 meters deep would have to evaporate $1000/(1.5 \times 17) = 39$ times, if we assume that the basin's water volume somehow could remain constant. Connection with the open ocean is needed to account for that refill potential. The notion that complete desiccation was rare is also supported by the fact that not every evaporite cycle goes all the way from gypsum to bittern salts. More commonly, the basin "sticks" for a long period of time in the salinity range of gypsum- or halite-deposition conditions. This indicates that inflow was large enough to feed new oceanic salt water into the basin, while small enough not to dilute the waters inside the basin by too much (in which case evaporite deposition would stop). That way, production of evaporites would continue for a long time in a basin that remained within a specific range of very high salinities. Sometimes, evaporative water loss might be a bit stronger than new inflow and the basin would concentrate a bit more. At other times, evaporative water loss would be a bit weaker than new inflow and basin salinity would reduce a bit. The observed evaporite types indicate the limits within which the basin salinities must have remained (notably, between about 140 and 350 for gypsum, and above 350 for halite).

Formation of salt giants was not a rarity in Earth's history; quite a few are known (figure 15). Some of the main salt giants are the Siberian salts and the Sverdrup basin salts (northernmost Canada) from the Cambrian period; the Moscow basin and the North American Elk Point basin salts of the Devonian; the Zechstein basin from the Permian, which covered most of northern Europe; the Gulf Coast basin in the Gulf of Mexico from the Jurassic; the South Atlantic salts from the Middle Cretaceous; and the Mediterranean–Red Sea–Persian Gulf evaporites from the Miocene epoch (for ages of these geologic intervals, see figure 2).

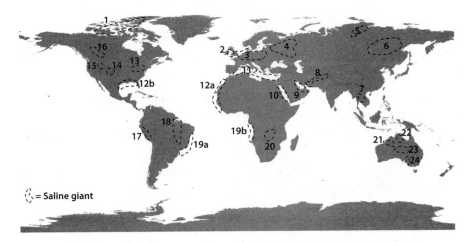

Figure 15. Map of major evaporite deposits (saline giants, or salt giants) of the past billion years or so, as found in the world today. Note that the continental configuration has changed greatly between the times these deposits were originally formed and the present, and that many older salt giants exist as well. 1, Sverdrup basin (about 330–300 Ma); 2, Northwich halite (245 Ma); 3, Zechstein basin (260–252 Ma); 4, Moscow basin (about 395 Ma); 5 and 6, Siberian salts (about 520 Ma); 7, Khorat basin (100–66 Ma); 8 and 9, Hormuz salt basin and salt range (about 545 Ma); 10, Red Sea evaporites (13–6 Ma); 11, Messinian Salinity Crisis (5.96–5.33 Ma); 12a and b, evaporites formed in the newly opening North Atlantic (170–160 Ma); 13, Salina basin (about 420 Ma); 14, Delaware basin (260–252 Ma); 15, Paradox basin (about 330–300 Ma); 16, Elk Point basin (388–383 Ma); 17, Permian salt in Andes (260–252 Ma); 18, Parnaiba and Parana basins (both about 330 Ma, and 118–113 Ma); 19a and b, evaporites formed in the newly opening South Atlantic (118–113 Ma); 20, Copperbelt evaporites (about 850 Ma); 21, Canning basin (about 445 Ma); 22, McArthur group (1750–1700 Ma); 23, Amadeus basin (about 520 Ma); 24, Flinders ranges (800–780 Ma). Ma stands for million years ago.

The evaporite deposits have found many economic uses, the most obvious of which is mining for the minerals themselves. Another commercially important aspect arises from the plastic "flow" behavior of salts under pressure, which causes salts to deform into pillars, domes, caps, and sheets. These structures form important reservoir traps and caps for petroleum and gas. And last but not least, evaporites are common targets for nuclear-waste disposal facilities, because of their geologic characteristics, predictable engineering and physical behavior, and sealed nature with respect to groundwater flow.

The Jurassic evaporites of the Gulf Coast basin date from about 170 to 160 million years ago. At that time, rifting started between North and South America as part of the continued breakup of the Pangaea supercontinent, during which the North Atlantic began to open as well. As we saw before, initial rifting of the Gulf region created a shallow basin that was connected to the Panthalassic Ocean on the outside of the C-shaped Pangaea supercontinent via a passage that had opened by separation of the North and South American plates. The subsiding crust was covered by almost four kilometers of virtually pure halite, over a time span of about 10 million years. This thickness and the lack of bittern salts are clear evidence for the existence of a connection with the open ocean. Eventually, more normal marine sediments formed over the evaporites. This marked a transition to less restricted conditions, owing to further opening of the basin, a sea-level rise, or both.

The South Atlantic evaporites date from the final 5 million years of the Aptian age, between roughly 118 and 113 million years ago. They formed in basins that resulted from the early phases of rifting between Africa and South America, which would continue to form the South Atlantic. These evaporites cover thick lake-sediment sequences that had started to form about 142 million years ago. So one might imagine that the earliest stage of rifting looked rather similar to the modern East African Great Rift Valley, filled with great lakes.

Soon after this rift valley connected with the open seas, likely at its southern end, evaporite deposition started in a restricted rifting basin with a width of about 300 to 500 kilometers and a length of more than 2000 kilometers. The evaporites built up to thicknesses of 1 to 2 kilometers. Eventually the basin became less restricted, and shallow marine sediments started to form over the evaporites. On this basis, we think that the basin in which these evaporites were deposited was—again—not a particularly deep basin.

*****

Here we need to step outside the general timeline of the book again. To get a deeper understanding of the potential controls on salt-giant formation, nothing beats a closer look at the most recent one: the Mediterranean salt giant, which formed throughout the entire Mediterranean basin of its time. It is the most extensively studied of all large salt giants.

The Mediterranean salt giant (see figure 15) has been very accurately dated by a variety of approaches, including magnetic-reversal dating, radiometric dating of volcanic ash layers, and the astronomical-tuning approach that relates sedimentary cycles to variations in the astronomical parameters. Thus we know that the Mediterranean salt giant formed over a period of 630,000 years, from 5.96 to 5.33 million years ago within the Messinian age at end of the Miocene epoch. As such, the event of its formation has become technically known as the Messinian Salinity Crisis (MSC). Although discussing it will again sidetrack us a bit from the chronological order of this book, the MSC is so well studied that it provides essential insights into how remarkable these salt-giant events really were.

The MSC started with deposition of the so-called lower evaporites, between 5.96 and 5.60 million years ago, mainly in marginal basins around the Mediterranean. The lower evaporites consist of an alternation between gypsum and marls deposited in the marginal basins. At the same time, cycles between oxygenated and euxinic marls were deposited in the deeper parts of the Mediterranean. Formation of the gypsum requires that enough new seawater entered the marginal basins to ensure a plentiful supply of new calcium and sulfate (Ca^{2+} and SO_4^{2-}) ions, that the water was concentrated to salinities of 140 or more (when gypsum starts to form), and that the water did not get sufficiently concentrated to start deposition of halite (NaCl), which would start at a salinity of 350. Also, given that the marginal basins cycled between gypsum and marine marl deposits, sea level clearly must have remained high throughout the lower evaporite phase.

We can thus infer that the Mediterranean was filled to a "normal" level with waters that—at least in the marginal basins—had become very salty, first because of restriction of exchange with open waters, and second because of the region's high rate of evaporation. Since gypsum was formed but not halite, salinities must have hovered between 140 and 350. In view of the alternation with nonevaporite marls, we can reasonably assume that salinities fluctuated closely around the lower gypsum limit; a drop below 140 then led to deposition of a marl, while a rise above that limit resulted in gypsum. The salinity fluctuations around 140 causing marl or gypsum deposition may have been caused by climate cycles between wetter and drier conditions, respectively. Alternatively, they may have resulted from minor fluctuations between higher and lower sea levels, which caused decreased or increased restriction of the basins, respectively. Or, of course, some combination of these controls may be possible. Opinions remain divided, and further research is needed to clarify the matter.

Then something exceptional happened between about 5.60 and 5.55 million years ago: the entire Mediterranean began to dry out. Its connections with the open ocean must have become very restricted indeed, so that inflow failed to compensate for the evaporative water loss. This restriction resulted from movement of the African tectonic plate against the European plate. Continental sediment deposits appear in the marginal basins, indicating that the sea was dropping fast and these basins became exposed. Trying to keep up with the drop in sea level, the Nile and other rivers such as the Rhone started to incise what would eventually become vast canyons, to depths of more than two kilometers below the present level.

Incidentally, the Nile gorge was discovered when Russian engineers were investigating the potential footings for the Aswan High Dam in Egypt. They expected to encounter the ancient, hard bedrock soon below the sediment surface, but instead encountered a very deep gorge filled in with river sands. This complicated the construction of the major dam, and at the same time delivered a tantalizing and unexpected view of the stunning potential scale of ocean variability.

At the time of the desiccation, major halite bodies formed in the deep Mediterranean that reached up to a kilometer thick, and locally more. Depositing all these salts only took an estimated 50,000 years. Especially in the eastern Mediterranean, the deposits also include bittern salts, which indicates that the diminishing waters at times became extremely concentrated, equivalent to salinities far above 400 parts per thousand. The great volume of halite deposited and the existence of cycles in the deposits in the eastern Mediterranean basin indicate that some (periodic) seawater supply from the open ocean must have persisted, despite the severe restriction of the ocean connections. Over time, sea level inside the Mediterranean basin dropped by hundreds of meters, and possibly even several kilometers.

The following episode, from 5.55 to 5.33 million years ago, was rather weird. Deposits from this episode are commonly referred to as the "upper evaporites." These formed under remarkable periodic water-salinity changes throughout the Mediterranean basin, as witnessed by cycles of gypsum beds of a type that forms in very shallow waters, and shales. Strontium isotope values in the gypsum bear witness to considerable freshwater input, and the intervening shales even contain rare and scattered brackish-water fauna. Toward the younger limit of this episode, the deposits show an even stronger freshwater signature, including fossil fauna and flora that suggest freshwater influx from basins to the north and northeast of the Mediterranean—modern remains of this so-called Paratethys environment are the Caspian Sea and Aral Sea. This unit is referred to as the Lago Mare, or "lake sea," unit, and it started in earnest at around 5.42 million years ago. Some gypsum deposits occasionally occur within the Lago Mare episode, but Paratethys fauna and flora remain represented throughout.

Certainly at the beginning of the upper-evaporite series, the Mediterranean seems to have been largely dry. In that case, the relationship of temperature with altitude in the atmosphere implies that, at two kilometers below sea level in the dried-out basin, surface temperatures would be 15°C to 20°C higher than at regular sea level, and thus may have been some 50°C to 60°C in summer. Recall also that this phase directly followed the major evaporite deposition. These salts would now have become directly exposed to the air. So the exposed seabed in the deep basins would have been very hot and replete with salt. Such an abundance of exposed salt would suck any moisture straight out of the air, and poison any rainfall or runoff puddles that collected. It is inconceivable that anything except extreme microbes could survive in such an environment.

During later Lago Mare stages, more sediment had arrived to cover the evaporites, and a more regular and steady supply of freshwater from the Paratethys basin diluted the salinity of any surface waters, allowing the (periodic) survival of brackish-water and even freshwater biota. But occasional recurrence of evaporite deposition indicates that conditions remained highly variable. For a modern analogue, it may be helpful to think about the Caspian Sea and Aral Sea, where conditions have also varied between arid episodes with evaporite deposition and wetter periods with lake-level rise and a return of low-salinity conditions. Lake Eyre in Australia may be a useful model too.

And then, abruptly, at around 5.33 million years ago, the Mediterranean was full with normal seawater again. We consider this to be the "moment" that the modern Strait of Gibraltar started to open. The abruptness of the switch from Lago Mare or continental deposits to open marine deposits suggests that this reconnection with the open ocean was not a gradual process on normal geologic time scales. Instead, the connection seems to have ruptured open abruptly. This would have caused a waterfall from the Atlantic Ocean into a deep, largely desiccated, Mediterranean basin. Recent work has estimated that 90% of the Mediterranean refill was accomplished within a few months to two years, causing sea level in the Mediterranean basin to rise by up to 10 meters per day. That would have been something amazing to behold, and the roar of the cascade must have been deafening.

By the way, to cut short any thought of relating this event to the flood legends of many cultures, remember that no humans were around 5.33 million years ago. Although the hominid lineage seems to have split from the chimpanzee and bonobo lineage by about 7 million years ago, it would take another 3 million years before the earliest bipedal hominids appeared with brain volumes that reached even a third of ours (*Australopithecus*). And it was only as recently as 2.7 or 3 million years ago that our first ancestors in the genus *Homo* appeared on the scene. So, sadly, no ancestors were around to witness what surely must have been the single most spectacular event in recent Earth history—such a shame. If ever a time machine is invented, then I put my hand up to go there and check it out—I'll be sure to bring some earplugs and a camera.

*****

To summarize, this chapter has taken us on a whirlwind tour through the Mesozoic greenhouse world, when the oceans finally settled into their modern, Cretan, biogeochemical mode of operation. It was a time of great warmth, but with major superimposed climate variations. The major end-Permian ocean-acidification event had provided an almost blank slate, because of major mass extinction, from which the Mesozoic started; the survivors rapidly diversified and radiated into the open spaces (or "niches," as we refer to them). Another large extinction event at about 201 million years ago, probably also associated with ocean acidification, created a further bottleneck effect that led to another phase of fast development and radiation within the Mesozoic. During this time, there was major carbonate deposition throughout the warm Mesozoic oceans, and the low-latitude Tethys Ocean was a prime location for it.

A major development through the Mesozoic was the breakup of the Pangaea supercontinent, along with the opening up of new, restricted, ocean basins. Large-scale external-carbon input from enormous volcanic episodes, generally high temperatures, and basin restriction conspired to cause anoxic events in the oceans on very large—in some cases possibly almost global—scales. Enormous quantities of organic carbon were buried in ocean sediments, and would over time mature to develop into some of the modern world's largest petroleum sources. We discussed more recent anoxic events in the Mediterranean Sea as minianalogues, to achieve a more detailed look at the typical processes involved. Finally, we evaluated the large Mesozoic deposits of sea salts, or evaporites, in the newly opening North and South Atlantic basins of the time. Again, we used a more recent analogue for that in the Mediterranean Sea, to get a more detailed feel for the typical controlling processes.

CHAPTER 7

WINTER IS COMING

Following the end of the age of reptiles at the Cretaceous-Paleogene-boundary extinction event of 66 million years ago, Earth warmed again. The Paleocene-Eocene Thermal Maximum (PETM) of about 56 million years ago was a relatively brief hot event atop this slower, long-term temperature rise toward a period known as the warm Early Eocene Climatic Optimum (EECO) (see figure 14). The Early Eocene warm period spanned from about 54 to 48 million years ago. The EECO peaked 52 to 50 million years ago, and was the warmest sustained period on Earth since the Middle Cretaceous. There was no ice on the planet, and the equator-to-pole gradient was much weaker than today. Crocodiles and palm trees appeared at high northern latitudes. Forests covered Antarctica. Tropical temperatures reached about 35°C and deep-sea temperatures up to 12°C. These deep-sea temperatures are a good indication that similar temperatures must have prevailed at the surface in winter at high latitudes. During summer, high-latitude temperatures in places exceeded 18°C.

Mammals started to radiate and diversify during the Paleocene epoch of 66 to 56 million years ago, and this especially accelerated during the PETM and the ensuing Eocene epoch of 56 to 33.9 million years ago. In fact, much of the flora and fauna that appeared during this time would be quite recognizable to us. But many massive reptiles remained. Examples include *Gigantophis* and *Titanoboa*, constrictor-type snakes of about 10 to 13 meters' length, respectively. Geographically nearly isolated, except for a connection to Antarctica at its southernmost extremity (which also separated about 31 million years ago), South America saw an arresting development of giant flightless terror birds that would roam the land alongside mammals until about 2 million years ago. In the oceans, sharks diversified rapidly, and the mammalian ancestor of whales and dolphins, *Ambulocetus*, appeared at around 50 million years ago.

This was a momentous period in terms of plate-tectonic rearrangements due to the ongoing breakup of Pangaea (see figure 4). Important ocean basins and passages opened up, and massive mountain ranges began to form. These would provide much fresh rock material for weathering.

On the South American plate, formation of the Andes has been ongoing since the earliest phases of breakup of the Pangaea supercontinent in the Triassic-Jurassic, but it was during the Cretaceous that the Andes became recognizable as they are today. Similar to the Rockies (see below), this resulted from intense subduction processes along the western margin of the plate. Mountain-building processes in the Andes continue strongly until the present, and include highly active volcanism.

Between about 80 and 55 million years ago, strong subduction processes at the western margin of the North American plate caused bunching up, buckling, and uplifting of massive packages of sediments that had formed within the plate, in and around inland basins such as the Cretaceous Western Interior Seaway. These deformations drove the formation of the Rocky Mountains. Initially, a highland plateau formed up to altitudes of about 6000 meters. Erosion has cut the current topography into that plateau over the past 60 million years or so.

Although the connection of the North Atlantic with the Arctic region had only started to open from about 60 million years ago as the North American and Eurasian plates reached full separation, the North and South Atlantic Oceans were opening fast during Paleocene-Eocene times.

Ever since the onset of the Cretaceous, successive subduction of Paleo-Tethys and Tethys—including the collision of Cimmeria with Eurasia—had caused large-scale crustal deformation that started the foundations of a mountain chain along almost the entire southern margin of Eurasia, from the Pyrenees and Atlas Mountains in the west, through the Alps, Apennines, Carpathians, Anatolian highlands, Caucasus, Zagros, Hindu Kush, Pamir, Karakoram, and finally the Himalayas in the east. The major phase of mountain building occurred between 66 and 2.5 million years ago, and it continues today in parts of this extensive range. By the time of the Eocene, the Tethys Ocean had already all but vanished between India and Eurasia, and it was closing fast between Africa and Eurasia. Finally, Australia started to separate from Antarctica from about 45 million years ago.

The EECO ended with the onset of a long, variable cooling trend. This onset, in turn, coincided in time with a phase of massive organic-matter deposition in the Arctic Ocean that was caused by enormous growth of a freshwater fern, called *Azolla*. The Arctic was nearly cut off from the open ocean at this time, which strongly limited the marine deep-water oxygenation potential. Ample river runoff from the surrounding continents created a freshwater layer on top of poorly circulating deeper seawater. We see similar—if less extreme—conditions in the Black Sea today. *Azolla* thrived in the fresh top layer.

The great volume of sinking and decomposing organic plant matter created strong oxygen demand in the deep waters. Because deep waters in the highly restricted basin were hardly circulating, the supply of oxygen was limited. As a result, the deep waters quickly became anoxic. *Azolla* is a fast-growing plant that gets its nitrogen from the atmosphere; its growth therefore depends only on the availability of enough phosphorus and trace nutrients in the waters where it grows. These are thought to have been in sufficient supply because of runoff from the surrounding continents, and potentially from phosphorus regeneration from anoxic sediments. In addition, sediment supply by the rivers efficiently buried the organic matter. The end result of the *Azolla* event (possibly not alone, but along with other carbon-removal processes), was a strong net carbon removal from the atmosphere-ocean-biosphere system. Atmospheric CO_2 concentrations dropped, within 800,000 years, from about 1400 ppm or more to below 1000 ppm. More extreme estimates may also be found in the literature, from 3500 to 650 ppm, but they don't find much support in more recent studies.

Occasionally after about 46 million years ago, and more notably after about 38 million years ago, centimeter-scale dropstones show up in marine sediments from the high northern latitudes. These pebbles are too large to have been transported by wind, and water couldn't have done it either in the relatively low-energy conditions of the open sea. The only way to import these dropstones is by freezing them into ice, which drifts out over the sea, where it melts and releases the stones. Therefore, there must have been ice by this time—most likely sea ice— that picked up rock grains at the coast. And this ice must have been substantial enough to survive transport into the open sea without immediately disintegrating, which is a clear indication that temperature, at least in winter, dropped below freezing for considerable periods of time.

In short, winter had come to the north by about 46 million years ago, hot on the heels of the major carbon removal at the time of the *Azolla* event. Quite soon, however, the general post-EECO cooling trend was interrupted by an up to 5°C warming at around 40 million years ago, known as the Middle Eocene Climatic Optimum, or MECO. That warming was associated with a CO_2 rise from about 750 to about 1250 or 1500 ppm. It has been attributed both to intensified volcanism and seafloor spreading, as Australia moved away from Antarctica after beginning its separation at around 45 million years ago, and to intensified volcanic degassing from rapid oceanic-crust subduction with large volumes of carbonate sediments as India closed in on Asia. From 40 million years ago, the overall cooling trend took over again. But still it would take until 14 million years ago before heavy perennial ice conditions appeared in the north.

Again based on dropstones in marine sediments, we know that winters with notable sea-ice activity had also developed around Antarctica by about 46 million years ago. Considerably later, at the Eocene-Oligocene boundary of 33.9 million years ago, the situation escalated in Antarctica. By that time a large ice sheet had become established on East Antarctica within about 300,000 years. Modeling studies suggest that it happened through merging of ice caps that first formed on mountain ranges, in a two-step process between 33.9 and 33.6 million years ago. These steps of glaciation were first recognized using stable oxygen isotope data from carbonate microfossils in marine-sediment cores, which offer a measure of global ice volume, as I will detail in the next section, Reconstructing Sea-Level Change. As soon as the glaciation became established, sea-ice conditions were

intensified, leading to seasonal productivity peaks capable of supporting krill masses and the major animals that feed on those krill masses.
A sharp increase took place in the numbers and sizes of large baleen
whale species.

The two steps of glaciation relate to particularly cold configurations of the astronomical climate cycles, against a background CO_2
drop from about 1000 to about 650 ppm. We still don't know exactly
why this drop happened, but some researchers point the finger at
intense weathering during a major mountain-building phase in the
Alpine and Himalayan ranges. At the same time, the Antarctic continent became more thermally isolated, as an early version of the Antarctic Circumpolar Current (ACC) appeared around 35 million years
ago owing to opening of the passage between Australia and Antarctica (full establishment of the ACC had to wait for deep opening of
the Drake Passage between South America and Antarctica, around
31 million years ago) (see figure 6). A strong ACC prevents penetration of warm currents to the Antarctic continent. Using a climate
model, the American researchers Rob DeConto and David Pollard
inferred that CO_2 lowering below 750 ppm was critical for allowing
the cold stages of orbital forcing to generate the ice sheet, while thermal isolation of Antarctica was of secondary importance. This fits
with a lack of precise timing agreement between the two glaciation
stages and the opening of the ACC passages.

The oldest ice dated in ice cores drilled from Antarctica is just over
one million years old. This is because ice flows like very thick syrup
from growth centers to the calving edges at the coast, so that it is continuously renewed. Although the ice itself is renewed that way, the
presence of a major ice mass on East Antarctica has been uninterrupted
since it first appeared almost 34 million years ago.

As in the north, therefore, winter also began to grip the southern
hemisphere (Antarctica) by about 46 million years ago. But—unlike
the north—it then managed to conquer the Antarctic continent as
early as 34 million years ago, establishing a deep freeze that has remained unbroken to the present day. The East Antarctic Ice Sheet
(EAIS) today holds water to an equivalent of 53 meters of global sea-
level rise. We would be wise to leave this sleeping giant well alone.

With the emplacement of the EAIS, the world had officially entered an icehouse state again, for the first time in more than 200 mil-

lion years. The presence of a significant, if variable, continental ice volume on the planet since about 34 million years ago has two important implications.

First, variations in continental ice volume cause relatively fast and large sea-level variations. Reconstructions show that changes in ice volume (sea level) and CO_2 forcing of climate have been systematically related all the while since 34 million years ago. On time scales of hundreds of thousands to millions of years, CO_2 changes through addition and removal of external carbon determined the longer-term glaciation state of the planet, and thus sea level. On time scales of thousands to tens of thousands of years, the two parameters have varied in concert in ways that may be expected from interdependent climate feedbacks. What I mean with this is that, on these time scales, polar temperature and associated sea-ice changes control the air-sea exchange of CO_2—which links to deep-sea CO_2 storage and release—and thus atmospheric CO_2 concentrations (see figure 8). The resultant atmospheric CO_2 variations in turn affect climate, including ice volume.

The second important implication of having a variable volume of ice on Earth concerns activation of the powerful positive ice-albedo feedback within the radiative energy balance of climate, which we discussed previously (see figure 10). Because of this feedback, any forcing of climate will have a disproportionately large impact at high latitudes. In other words, polar climate changes become amplified relative to global climate changes. This amplifies the interdependence between ice cover and CO_2 variations outlined in the previous paragraph.

The total continental ice volume during the Oligocene epoch (33.9 to 23 million years ago) varied between warm extremes, with 50% less ice than the volume of the modern EAIS, and cold extremes, with 25% more ice than the modern EAIS. It was long thought that there was a Late Oligocene period of warming, but it was later found to be an artifact in early temperature reconstructions. In contrast, up-to-date detailed records recognize two major intervals of glaciation in the Late Oligocene, at around 24.4 and 23 million years ago. After these, climate stayed generally cool until a relatively warm period called the middle Miocene Climatic Optimum (MCO), 17 to 15 million years ago.

CO_2 levels were about 350 to 400 ppm during the MCO, compared with 200 to 260 ppm before and after it. Temperatures at the Antarctic coast went up to about 7°C, and tundra developed around the margins of the ice sheet. Such temperatures are 11°C higher than today in the region, but—as explained above—polar responses to climate change had become amplified with the appearance of ice on the planet. Global temperature was at most 5°C higher during the MCO than today.

At around 15 million years ago, the end of the MCO coincided with a sharp drop in CO_2 levels by some 100 or 150 ppm. This has been attributed to a phase of increased uplift, weathering, and erosion in the Himalayas, and there was also a major phase of organic-carbon burial in the Monterey Formation of California, which today is a considerable source of oil. A variable ice sheet appeared in West Antarctica, and the EAIS reached its full proportions. There were ice-volume fluctuations after that time, but—truth be told—the big news item after this phase of "filling up" of Antarctica really is the development of continental ice volumes on the northern hemisphere, which happened much later, at around 3.3 million years ago.

Before launching into that discussion, we need a bit more detail about how long time series of past sea-level changes may be measured, as these give us the best clues about the amount of ice volume on the planet, and thus about ice-age climate cycles. Thereafter, we pick up the narrative again with discussion of the last few million years, when northern-hemisphere ice ages became increasingly important.

RECONSTRUCTING SEA-LEVEL CHANGE

Sea-level fluctuations over periods up to hundreds of millions of years are reconstructed with a method that was developed in close collaboration with the petroleum industry. In 1977, an Exxon team led by the American geoscientist Peter Vail produced a first record for the past 550 million years. It was significantly updated in 1987–88 by the oceanographer Bilal Haq. Since then, updates have especially focused on improving age control and details in specific intervals of interest.

This body of work comprehensively demonstrated that sea level stood up to several hundreds of meters above the present level for much of the past 550 million years. We have seen before that this was a consequence of plate-tectonic processes. Superimposed on those long-term, slow cycles, many faster cycles of shorter duration and smaller amplitude have been found. These mainly relate to continental ice-volume variations, the focus of this chapter, and resolving them in sufficient detail requires other methods.

For determining detailed records of past sea-level variations, fossil corals are a convenient choice. Coral reefs are well known for developing close to sea level, and corals are widely distributed in tropical to subtropical regions around the world, so there are many islands and other coastlines to investigate. Best of all, corals build a skeleton of aragonite (a type of calcium carbonate), which is very suitable for radiometric dating, using either radiocarbon or uranium-series techniques.

The coral method in its most basic form is straightforward. A fossil coral deposit is identified and sampled. The current altitude (or depth if underwater) of the sample is carefully measured relative to modern sea level. Next, the sample is checked for suitability, and radiometrically dated. This gives us the modern position of the sample and its age; all we need now is to correct for the amount of uplift or subsidence of our sample location since the time the coral was growing. There are different ways of doing so, and all include unavoidable and considerable uncertainties that need to be dealt with. Once a rate is determined, it gives us the vertical displacement of our sample location, which we can combine with its age and current altitude (or depth), to work out at what position relative to modern sea level our sample was growing originally. Past sea level then may be determined by taking into account the common living depth of the species investigated.

In reality, many other corrections and adjustments are involved; the above procedure outlines only the fundamental underlying principles. There are many sources of uncertainty as well, and these are dealt with by expressing past sea-level values with a band of well-quantified uncertainties. But by taking many samples, from many locations in many different places around the world, a picture eventually emerges of where sea level stood at different times in the past. Currently, there are about 3000 to 4000 well-dated coral-based values for past sea level, spanning the last 300,000 years or so. But most of those data are focused on the last 30,000 years, and on the interglacial of 130,000 to 110,000 years ago when sea level stood 5 to 10 meters higher than today.

In addition to fossil coral reefs, there are also other types of shallow marine or coastal landforms that can be used as sea-level markers. These include beach deposits, drowning surfaces, and cave deposits that are active only when the cave is above sea level. The controls on these are obviously different from those on reefs, but the general way in which the data are used is fairly similar.

Another way of evaluating past global ice-volume changes, and thus sea-level changes, relies on deep-sea oxygen isotope ratios. Earth contains three stable isotopes of oxygen: oxygen-16, oxygen-17, and oxygen-18. They all have 8 protons in their cores, which is what makes them oxygen, but different numbers of neutrons (8, 9, or 10, respectively). The most abundant isotopes are oxygen-16 and oxygen-18 (we will ignore the very rare oxygen-17 isotope here). The slight mass difference between these, due to the two extra neutrons in the atomic core of oxygen-18, causes a slight difference in behavior of molecules containing either isotope during chemical reactions.

During evaporation, oxygen-16 is favored in the water-vapor phase, and relatively more oxygen-18 remains in the sea. Upon condensation, to form rain, oxygen-18 is favored, so that rain contains relatively more oxygen-18 than the original vapor. The removal of this rain leaves the remaining vapor relatively enriched in oxygen-16. Eventually, when the last vapor precipitates out (as snow) at high latitudes, it contains a strong enrichment in oxygen-16. As this is stored for many tens to hundreds of thousands of years in an ice sheet, the net removal of this mass of oxygen-16 causes a proportional enrichment in oxygen-18 in the oceans—the more ice volume increases, the more the oceans are enriched in oxygen-18. We can measure this by changes in the ratio of oxygen-18 to oxygen-16 in seawater, relative to an international standard. We call this the delta-^{18}O ratio. Thus, the delta-^{18}O of seawater is a measure for global ice volume.

From sediment cores, we cannot directly measure the delta-^{18}O of seawater, but instead we measure the delta-^{18}O of carbonate microfossil shells. This leads to a complication, as the delta-^{18}O of carbonate is a function of both the delta-^{18}O of seawater and the seawater temperature. So we need to correct for that temperature influence. We do so by measuring the ratio of magnesium to calcium (Mg/Ca) in exactly the same shells, which is temperature sensitive. Correction for the temperature influence leaves us with the delta-^{18}O of past seawater. Note, again, that this description represents only the underlying principles of the method.

Delta-^{18}O records for carbonate deep-sea microfossils have grown into the backbone of paleoceanography, driven especially by insights into their utility that hark back to the work of Cesare Emiliani in the United States and Sir Nicholas Shackleton in England. Decades of effort in the wider community of researchers have been compiled by the American paleoceanographers Ken Miller, Jim Zachos, and Maureen Raymo into detailed continuous records that span the past 65 million years. Care is needed when interpreting these records, as they concern delta-^{18}O of carbonate and therefore reflect a combination of ice volume and temperature. Although temperature variations are much smaller in the deep sea than in surface waters, they cannot be ignored. To resolve this, many researchers now make microfossil Mg/Ca-based temperature corrections, which were pioneered as routine measurements by Dirk Nürnberg from Germany, David Lea and Howie Spero from the United States, and Harry Elderfield from England. As yet, however, continuous and highly resolved temperature-corrected delta-^{18}O records span only a few million years. A few compilations have suggested extensions to 50 or 100 million years ago, but changes in ocean Mg/Ca ratios need to be accounted for over such time scales, and this is not straightforward. So we have a long way to go; it involves a massive analytical effort.

Once a temperature-corrected record for delta-^{18}O of seawater is established, we can relate it to ice-volume changes in different ways. One approach to doing so is via calculations or modeling, and the other is through calibration between changes in delta-^{18}O of seawater and independent sea-level data, notably from corals. Both approaches have their own assumptions and uncertainties, but we can add these to other uncertainties in the overall method, and plot the combined uncertainties around our reconstructed records of ice volume or sea level, to indicate how good an estimate at any point of time they may be. Unfortunately, there are certain issues that can upset the relationship between delta-^{18}O of seawater and sea level. For example, large (partly) floating ice shelves that extend into the sea from continental ice sheets displace almost their own volume in water. Their melting therefore hardly affects sea level, while their ice volume still strongly affects delta-^{18}O. A sound understanding of such problems is needed to support correct interpretation of seawater delta-^{18}O.

Note that, for times before about 34 to 40 million years ago, there was no significant ice volume on the planet, so that the entire delta-^{18}O signal is commonly considered to reflect only deep-sea temperature changes. That assumption underlies a lot of the deep-sea temperature estimates that you have seen cited in this book.

In contrast to the fossil-coral method, the delta-^{18}O method does not lend itself to high-precision dating. But it does produce beautifully detailed and continuous records that extend over millions to tens of millions of years. Overall, dating is not too bad either, as the delta-^{18}O variability can be mathematically related to the astronomical cycles of climate forcing.

There is a third method of determining past sea-level variability, which I have pioneered with my own research group. Starting in the mid-1990s, we began using delta-^{18}O records from carbonate shells of surface-dwelling microzooplankton in evaporative marginal seas that have only narrow and shallow straits connecting them with the open ocean. We specifically targeted the Red Sea and Mediterranean because—at 137 and 284 meters' depth, respectively—the straits that connect these seas to the open ocean are very shallow relative to the glacial-to-interglacial range of sea-level change, which is between about 130 meters below and 10 meters above the present level. In addition, the basins are highly evaporative, and have been so for as long as we know.

The size of a strait limits the amount of water that can go in and out of a basin. We calculated this for different sea-level positions and evaporative forcing, using so-called hydraulic-control models. As the strait becomes smaller, notably due to a sea-level drop, less water can be exchanged through it. As a result, water then resides longer inside the evaporative basin, in a similar way as when you partially block the plughole when draining a bath. In consequence, the water is exposed

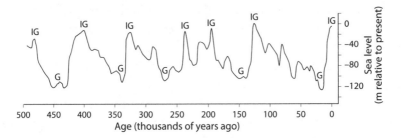

Figure 16. Sea-level history for the past 500,000 years. This reconstruction derives from the Red Sea method; IG denotes interglacials, and G indicates glacials.

for longer to evaporation, causing more oxygen-16 removal by that process. This, in turn, causes the seawater to become more enriched in oxygen-18. In short, the lower the sea level, the more oxygen-18-enriched the water (and thus the shells formed in that water).

For both basins, we determined the changes in delta-^{18}O of carbonate in relation to sea-level change in the straits. In the open ocean, glacial-interglacial delta-^{18}O changes span about 1.5 parts per thousand. In the Mediterranean, the residence-time effect increases this to about 2.5 parts per thousand, and in the very restricted Red Sea it is almost 6 parts per thousand. So the Red Sea gives the best sea-level estimates with the smallest uncertainties.

Finally, the calculated sensitivities of delta-^{18}O in the basins to sea-level change in the strait are used to determine sea-level records from long, continuous, and detailed microfossil delta-^{18}O records from sediment cores. The longest useful cores from the Red Sea span 500,000 years. Although the method has larger uncertainties in the Mediterranean, there is a payoff of a different kind: detailed, continuous records from that basin go back all the way to 5.33 million years ago.

Because there is no direct radiometric age control before 40,000 years ago (the practical limit of the radiocarbon method), dating is not as good in the marginal-seas method as in the coral method. But it is better than in the deep-sea delta-^{18}O method because Mediterranean sediment records can be more tightly linked to astronomical cycles, based on monsoon (sapropel) and dust changes that directly relate to precession, as discussed previously. In addition, the last few hundred thousand years of the Mediterranean and Red Sea records show close relationships with stable oxygen isotope records from cave stalagmites, which have been very precisely radiometrically dated. Overall, the method has thus provided a continuous record of sea-level change over the past 500,000 years, with good age control (figure 16).

By comparing reconstructions from several different independent methods—all with their own issues—a common signal can be identified that indicates the most likely sea level at any point of time in the past.

THE GREAT NORTHERN ICE AGES

We previously discussed that a climate model suggested a CO_2 threshold of 750 ppm for Antarctic glaciation. The same model also suggested a CO_2 threshold of 280 ppm for major continental-scale glaciation on the northern hemisphere. A recent compilation of past CO_2 changes suggests that this level was first approached 14 to 15 million years ago. Observations of coarser material transported by ice, like dropstones, suggest that glaciation on the northern hemisphere started at around 10 million years ago. This included initial formation of a minor ice sheet on southern Greenland, which eventually expanded 3.3 million years ago. Thus we arrive within the time frame of the most recent 3.5 million years, which spans the most profound changes in the history of northern-hemisphere glaciation.

Between 3.5 and 2.5 million years ago, sea level varied between extremes of about 10 to 30 meters above the present level, and CO_2 reached up to 400 ppm. Ice-age fluctuations began to include periodic glaciation outside Greenland by 2.8 million years ago, in the Eurasian Arctic, northeastern Asia, and Alaska. At around the same time, a widespread cooling took place, along with a CO_2 drop to values below 280 ppm, which has been attributed to increased storage of carbon in the deep ocean.

The first significant glaciation on the North American continent dates back to about 2.5 million years ago. Roughly around that time, northern ice-volume growth intensified to a point that sea level recorded its first drop to more than 70 meters below the present. This is more than half the drop seen during the full-fledged ice ages of the past 500,000 years. It is worth pausing a little to reflect on this. A drop to 70 meters below the present sea level implies that Earth contained roughly twice as much ice as it does today. All of Antarctica today contains ice to an equivalent of about 58 meters of sea-level rise (and Greenland about 7 meters). Also, Antarctica is almost full of ice; it is estimated that Antarctica could—at the very most—only take an extra 15 meters' sea-level equivalent of ice volume. In a period with 70 meters' sea-level lowering relative to the present, this implies that the northern hemisphere contained an ice mass worth at least 55 meters of sea level, or almost the same size as the modern Antarctic ice sheet. That ice was mostly located on northern Europe and Asia, and North America. As more severe ice ages developed, this extra mass grew even larger.

Although I work on this topic every day, I remain totally in awe that the northern hemisphere, which today holds only the relatively small Greenland ice sheet, could have contained between one and eventually two Antarcticas' worth of extra ice volume during the ice ages since 2.2 million years ago. Even more remarkably, this massive volume was supersensitive to climate change, given that it built up and completely disappeared with great regularity. Such behavior is in stark contrast with Antarctica, where great ice mass has been a fixture since 34 million years ago.

The main trend of increasing ice volume appears to relate to a slow, long-term lowering of CO_2 concentrations, which preconditioned the climate system sufficiently for astronomical cycles to trigger glaciations. Once glaciations started, they were relentlessly pushed along by temperature and sea-ice controls on CO_2 exchange with the ocean and thus with the massive deep-sea reservoir, and by strong climate feedbacks that include ice- and snow-albedo feedback, vegetation feedback, and dust-aerosol feedback. The development of large ice sheets had such profound implications that the feedbacks became more important for the radiative forcing of climate than the initial astronomical cycles. For example, the feedbacks over the last five ice-age cycles caused at least seven times more change in the annually averaged energy contrasts over Earth than the initial astronomical forcing.

By about 2.8 to 2.2 million years ago, something must have happened to cause cooling, along with a drop in CO_2 levels that may reflect increased deep-ocean carbon storage, and eventually the first of many major ice ages on the northern hemisphere. There are several hypotheses for this. A popular family of concepts involves some consequence of closure of the Panama Isthmus, the thin land bridge whose emergence closed a passage between the Atlantic and Pacific Oceans, which changed the amount of oceanic poleward heat transport. However, explanations that rely on this closure lost some traction when it emerged that the closure started much earlier, before five million years ago. The most recent findings, based on fossils and sedimentary evidence that indicate an environment above sea level, place the start of closure as early as 15 to 12 million years ago.

It may be that the closure hypothesis is still valid, and that it relates to the final closure of a few remaining (shallow) marine passages. If not, then a new concept is needed; there are plenty of alternative hypotheses to play with, from changes in ice-sheet dynamics to reduction in CO_2. Several theories ascribe the latter to continental weathering, increased high-latitude ocean stratification, increased biological-pump efficiency, or changes in the tropical ocean. And, of course, combinations of the various processes cannot be excluded either.

Clearly, the major question of what drove the onset of northern-hemisphere glaciation will demand much attention before it may be settled with confidence. It's a perfect mystery—join us if you would like to try to solve it!

Ice-age cycles remained tightly linked to the 41,000-year astronomical obliquity cycle until about 1.2 to 0.8 million years ago. At that time, the so-called Mid-Pleistocene Transition (MPT) led to ice-age cycles with an average spacing of about 100,000 years, for reasons that we don't really understand yet (another perfect mystery). The 41,000-year ice-age cycles before the MPT were symmetrical in shape, while the 100,000-year cycles after the MPT were saw-toothed, with gradual buildup of ice sheets over tens of thousands of years and rapid decay over thousands of years.

These underlying patterns over thousands to tens of thousands of years, however, have been overprinted by strong variability on time scales of centuries to millennia. Observations have revealed a long trend of ice-volume buildup since the end of the last interglacial, roughly 115,000 years ago. This trend had strong superimposed shorter-term sea-level variability, most notably in the form of ups and downs of about 30 meters over time scales of several millennia. Eventually, an ice-volume maximum was reached that marked what we refer to as the Last Glacial Maximum (the last ice age) of 25,000 to 19,000 years ago. This maximum ended with a fast ice-volume reduction and thus sea-level rise; by about 6000 years ago, sea level had jumped back up to about its modern position. Thus, the last glacial cycle shows a saw-toothed shape that is typical of ice-age cycles since the MPT. It is only in the detailed records of the past half million years that we can measure the rates of rapid ice-volume changes in the past, under natural conditions of climate forcing, as well as the durations of sea-level rise from start to peak rates. Such observations can be compared with models that apply our best—but always incomplete—knowledge of ice physics to estimate past ice-volume change, to determine which process representations in the models best approach the observed reality. Once models have been thus honed to best represent past natural sea-level variations, they can be used with more confidence for making sea-level change projections into the future, based on the more extreme, faster, human-induced climate forcing.

The first detailed insight into the rates and adjustment time scales of sea-level rise was presented in 1989 by the American paleoceanographer Richard Fairbanks, based on fossil coral data for Barbados. These data concerned the rapid global deglaciation from 19,000 to 6,000 years ago, which saw a global sea-level rise from about 130 meters lower than today toward its present-day position. This implies an average rate of rise of about one meter per century, which was not such a new finding. But Fairbanks also found that there were brief interludes within the deglaciation with much faster rise (up to four or five meters per century) in between longer intervals of slower rise. Such rapid jumps presented very new and rather alarming information about the operation of the climate system. Unfortunately, this was but one set of observations, for one specific interval of time. For a representative view of what is possible under a range of climate conditions, we needed to consider more events.

Assessments using the Red Sea method of sea-level reconstruction have provided a sea-level signal both in sufficient detail and with sufficient age control, to determine meaningful rates of sea-level change for many dozens of events. A first observation is that rates of sea-level lowering, related to growth of continental ice, were limited to a maximum of about one meter per century. This makes good sense when compared with calculations based on the surface area of ice-sheet regions and the available water-equivalent of average snowfall. It is hard to conceive of mechanisms that would make ice grow much faster than this. With respect to sea-level rise, the Red Sea record characterizes some 120 major events within the last half million years. Occasionally, the rates of rise were found to reach as high as five or even six meters per century. Such extreme rates of rise were found only at the ends of major ice ages, when the rises started off from a period with two to three times the modern ice volume. Sea-level rises that followed periods with less than two times the modern ice volume were mostly limited to a maximum of one to two meters per century.

The Red Sea records also show that the time needed from the start of a pulse of sea-level rise to its maximum values is measured in centuries. It happened within 400 years in 90 out of the 120 cases investigated, and within about 1000 years for nearly all the rest. In other words, continental ice sheets may have great mass, but this does not make them nonresponsive laggards. Instead, they appear to commonly respond within centuries, and to reach maximum retreat velocities well within 1000 years. Given that human-induced warming has now been going on for 150 to 200 years, we are getting within the "zone" in which significant ice-volume responses should be expected even if everything operated on slow, natural time scales. We ought to expect even stronger responses if we factor in today's exceptional human-induced rate of change in the energy balance of climate.

OCEAN CONTROLS ON CO_2

Prior to a discussion of the mechanisms of CO_2 variation during the last couple of ice-age cycles, we should first spend some time on ice-core studies. Ice-core studies have provided highly detailed CO_2 records for the past 800,000 years (top panel of figure 11), along with a range of other climate parameters. In consequence, the development of methods to drill continuous ice cores out of the great ice sheets has transformed our understanding of climate variability during the past several hundred thousand years.

Some of the earliest research on the great ice sheets dates from the 1930s, with investigations of vertical structures in the walls of hand-dug holes, or "ice pits," some of which reached tens of meters in depth. The concept of using ice cores for studying past climates arose in the late 1940s, and was strongly promoted from the early 1950s by the Danish researcher Willi Dansgaard. Dansgaard demonstrated this application using an ice core that was drilled through 1390 meters from top to bottom of the ice sheet, at Camp Century on Greenland. Next, Dansgaard and others drilled the first purely scientific ice core between 1979 and 1981 at the DYE-3 station on southern Greenland. Many ice cores have since been drilled on Greenland by American and European teams, and they span about 120,000 years of uninterrupted ice. The most notable examples are the efforts by the American Greenland Ice Sheet Project 2 (GISP2, where GISP1 was the drilling at DYE-3), the European Greenland Ice Core Project (GRIP), the broadly international North Greenland Ice Core Project (NGRIP or NorthGRIP), and the North Greenland Eemian Ice Drilling (NEEM) project.

Between the 1970s and 1996, the Soviets/Russians drilled long ice cores at the coldest place on Earth, Vostok station on Antarctica. This ice-core record reached a total of 3623 meters down, and spans the last 415,000 years in an undisturbed manner. In 2004, the European Project for Ice Coring in Antarctica (EPICA) completed recovery of a core at an Antarctic station 560 kilometers away from Vostok, which extended the time span covered by Antarctic ice cores to the last 800,000 years.

Ice cores are like sediment cores from the seafloor in that the youngest material resides at the top and the oldest material at the bottom, but they have three distinct advantages. First, ice (snow) accumulates much faster than seafloor sediment, so ice cores offer the opportunity for research in much smaller steps of time. Second, ice accumulates as snow in seasonally distinct layers, and the resultant ice banding can often be counted down along a considerable portion of the cores, which offers great dating potential. Third, ice cores contain two information carriers; namely, the ice itself and the air bubbles of ancient atmosphere trapped within the ice. Both can be analyzed in high detail, and—with some adjustments—information from both can be expressed on one and the same age scale.

The latter in particular revolutionized paleoclimate research. Air bubbles in ice have allowed direct measurement of past fluctuations in ancient CO_2 and CH_4 levels, along with a host of other gases and their isotopic compositions. Meanwhile, stable oxygen and hydrogen isotope ratios of the ice itself are good measures of past polar temperature changes. Finally, concentrations of dissolved and particulate windblown dust have also been measured using samples of the ice itself. In consequence, ice-core records have provided detailed, well-dated templates of variability in a host of critical climate parameters. They have grown into cornerstones of research into past climate variability.

CO_2 levels measured at polar latitudes are a good indication of global CO_2 variations because the gas is well mixed through the atmosphere. Detailed ice-core records of temperature and CO_2 levels demonstrate that these two parameters have strongly covaried during the ice-age cycles of the past 800,000 years. This is because temperature and CO_2 levels develop together in a tightly connected feedback cycle—CO_2 influences temperature, and temperature influences processes of CO_2 uptake or release, both with respect to the biosphere and the oceans.

Here, I should emphasize that the timing and intensity of feedback responses depend on the time scales of all processes involved. As a result, we may expect some timing offsets between temperature and CO_2 variations. Sometimes one changes a bit earlier than the other, sometimes it's the other way around, and at yet other times there seems to be no offset. Such offsets mostly remain limited to a few centuries, but may at times increase beyond that. This happens, for example, when a large and fast surface change triggers a feedback response that involves slower deep-ocean processes, or when other long-term feedback processes are activated as well, such as the ice-albedo and vegetation-albedo feedbacks. When observed in the records, such offsets are sometimes used (by people aiming to cast doubt on greenhouse-gas impacts upon temperature) to argue that temperature and CO_2 changes are not linked, and/or that temperature change leads and thus allegedly causes CO_2 change, rather than the other way around. In reality, such offsets merely reflect the complex and interrelated nature of the processes behind climate change—the total response represents the net influence of the initial disturbance and other associated changes, integrated over a longer period of time.

Overall, the ice-core records show that CO_2 typically varied between about 180 ppm in the most intense ice-age maxima to 280 or 300 ppm in interglacials (see figure 11). So we have about 100 ppm of CO_2 change to explain between glacials and interglacials. The ice-core records show important CH_4 contrasts as well, between about 350 parts per billion (ppb) in glacials and about 700 ppb in interglacials. The CH_4 variations were important in terms of greenhouse properties, as CH_4 has about a 30 times stronger greenhouse effect than CO_2. But in terms of amounts of carbon, the CH_4 variations were not so important, as CH_4 concentrations are 500 to 1000 times smaller than CO_2 concentrations. When considering movements of carbon through the climate system, we can focus on just CO_2. Figure 8 provides useful support to the following discussion.

During an ice age, major continental ice masses are formed on the northern hemisphere, surrounded by large permafrost regions—zones that are almost barren with respect to vegetation or occupied by tundra. These surface covers occupied mid- to high latitudes, replacing the dense forests, bogs, and swamps of interglacial times. As a result, there was much less carbon stored in the terrestrial biosphere during the cold glacials than during the warm interglacial periods in between. Because matter doesn't just magically vanish into nothing, or appear out of nothing, this means that portions of the carbon that were no longer in the biosphere need to have been in the oceans and/or atmosphere. If anything, therefore, we might expect the contracted glacial biosphere to result in higher glacial atmospheric CO_2 levels, but this is the opposite of what we observe.

Where was the carbon? Given that it was not in the atmosphere, and not in the biosphere, it must have been in the oceans—all 100 ppm's worth atmospheric carbon, plus the carbon involved in the glacial biosphere contraction. This adds up to 200 GtC from the atmosphere, plus 500 GtC from the biosphere. So we're talking about

700 GtC. To understand the meaning of this number, let's try to visualize it. Carbon in the form of the pure graphite of an artist's pencil weighs 2250 kilograms per cubic meter, so that 700 GtC represents 311 billion cubic meters of graphite, which could coat the entire land surface area of Earth in a two-millimeter-thick layer. Or we can view it differently, as a column that connects Earth with the Moon, which is 384,400,000 meters away. The amount of 700 GtC then represents a round column of pure graphite with a diameter of 32 meters, all the way from Earth to the Moon. That is a lot of carbon—and all of it moved into the oceans as ice-age conditions developed, and out again as ice-age conditions terminated. The Earth system is amazing at shuffling carbon around, and it's worth considering the processes behind it. First, however, remember the depth of time; nature used many thousands of years to shift this amount of carbon. In consequence, the volume of these adjustments may be mindboggling, but their rate—about 700 GtC in about 10,000 years, or 0.07 GtC per year—still was 30 or more times slower than the human-induced net carbon emissions of about 420 GtC in 150 to 200 years since the onset of the industrial revolution (an average rate of 2.1 to 2.8 GtC per year).

Solubility of CO_2 into the oceans increases with lowering of temperature, and just the ice-age drop in temperature by itself can explain about 20 to 30 ppm of the CO_2 drop from interglacials to glacials. But this drop is more or less offset by the CO_2 rise caused by transport of land-biosphere carbon into the oceans, along with the atmosphere-ocean gas-exchange impacts of increased ocean salinity and decreased ocean volume. This leaves almost 100 ppm of atmospheric CO_2 storage in the oceans to be explained by means of other processes on the interface between the organic and inorganic carbon cycles, and circulation. To explain the atmospheric CO_2 changes, including biosphere carbon changes via the atmosphere, we have to think about *net* changes in the exchange with the ocean. That means that we have to consider the balance between CO_2 uptake into the oceans and CO_2 release from the oceans.

Today, a large part of the CO_2 uptake into the oceans happens during the formation of deep water, and most notably the formation of North Atlantic Deep Water in the northernmost North Atlantic.

There are two main elements to this. First, the greater the volume of new NADW formation, the more CO_2 will be taken up into the deep ocean. Second, the more CO_2 has entered the surface waters before they are transformed into new NADW, the more CO_2 will be taken up into the deep ocean. If we assume that everything else remained constant, then an increase in CO_2 uptake into the oceans will cause a decrease in the atmospheric CO_2 level, until the air-sea contrast becomes small enough that a new equilibrium is established. Conversely, a decrease in CO_2 uptake into the oceans will cause an increase in the atmospheric CO_2 level.

NADW formation rates have fluctuated because of changes in the net freshwater loss or gain from the North Atlantic. During periods of strong evaporative water loss from the basin, NADW formation was stronger than during periods with net freshwater input, for instance due to meltwater input from ice sheets around the basin. Meanwhile, increased CO_2 concentrations in the surface source waters for NADW could result from such processes as reductions in temperature, increases in windiness (rougher water), and lengthening of the duration of surface-water exposure to the atmosphere, allowing gas exchange over longer periods of time.

CO_2 release from the oceans into the atmosphere today predominantly happens in the Southern Ocean. There are strong indications that this "leak" was stemmed during glacial times, due to sea-ice expansion, reduction of vertical mixing, and deep-sea circulation changes. The net CO_2 exchange in the Southern Ocean may even have been reversed during these times, based on "iron fertilization" due to wind-blown dust over the Southern Ocean, which caused increased productivity and biological-pump action, resulting in net CO_2 removal from the atmosphere into the deep sea. Over thousands of years, the extra CO_2 in the deep sea caused carbonate dissolution, moving the CCD to a shallower level. In addition, circulation changes in the deep sea, characterized by a stronger contribution of Antarctic Bottom Water, increased this carbonate-dissolution influence. And a third effect, through more intricate carbonate chemistry, caused a reduction of coral-reef formation during glacials that drove an increase in CO_2 solubility in seawater. These key influences on CO_2 storage in the deep oceans are thought to account for most of the drop in atmospheric CO_2 levels during glacials. But the details and contributions of each effect remain strongly debated.

During the rapid (in geologic terms) ends of ice ages, or deglaciations, the ice volume on Earth shrank so quickly that sea-level rises of up to 130 meters happened over little more than 10,000 years. But deglaciations were not all straightforward shifts from glacial to interglacial climate conditions. Most contained one or more climate swings between very cold and much warmer intervals. Ice and marine core records reveal that the timing of such alternations was slightly offset between the northern and southern hemispheres, in a "seesaw" manner, which we'll discuss in the next section, A Seesaw in the Ocean. These seesaws indicate that the climate swings within deglaciations primarily represent events of heat redistribution between the hemispheres, rather than global warming and cooling. The broader underlying longer shift from glacial to interglacial climate through each deglaciation, however, did represent global warming. In other words, each deglaciation represents a global warming, by about 5°C (see figure 14), with superimposed events of heat redistribution back and forth between the northern and southern hemispheres, on time scales of a few thousand years.

Also with some superimposed variability, atmospheric CO_2 levels markedly increased through the deglaciations, from glacial levels of about 180 ppm to natural interglacial levels of about 280 ppm. At the same time, the biosphere on land expanded. The total return of carbon to the atmosphere and biosphere over a 10,000- to 13,000-year deglaciation is similar to the amount locked away during buildup of ice ages, namely a total of about 700 GtC.

The concepts for getting the carbon out of the oceans are the reverse of the concepts for locking the carbon into the oceans that we discussed above. Hence, the idea is that initial warming due to orbital insolation changes led to decreased CO_2 uptake in the northernmost North Atlantic. In addition, release of CO_2 out of the oceans into the atmosphere was reestablished, especially in the Southern Ocean, in response to changes in the deep-sea circulation, increased vertical mixing, and reductions in sea-ice cover. As atmospheric CO_2 increase began, this drove further warming and circulation responses through a complex suite of feedback processes. Moreover, any iron fertilization due to windblown dust over the Southern Ocean would have been reduced toward levels similar to those seen today, as the ice age ended and sea level rose. Finally, sea-level rise flooded shallow regions, giving rise to a strong expansion of coral-reef habitats, and the resultant increase in coral-reef formation drove a decrease in CO_2 solubility in seawater. Again, the details, contributions, and exact timing relationships of the various influences remain debated.

A SEESAW IN THE OCEAN

Above, I mentioned heat-redistribution events between the northern and southern hemispheres within deglaciations. Such events, on time scales of a few thousand years, have also been found within ice-age periods. They have become descriptively known as bipolar temperature-seesaw events.

The bipolar temperature seesaw was first noticed using ice-core records from Greenland and Antarctica, and has since been found well expressed in many marine-sediment-core records as well. The initial discovery came from ice-core analyses in both polar regions. These provided temperature reconstructions based on stable oxygen or hydrogen isotopes of the ice itself, and atmospheric gas concentrations (including CH_4) using air bubbles contained with the ice. Thus, with some corrections, detailed coregistered records, measured on the same samples, were acquired of temperature and CH_4 levels.

CH_4 is rapidly mixed throughout the atmosphere within a matter of years, so that CH_4 variations are very similar between the two polar regions. This provided a great opportunity: CH_4 records could be used to relate the timing of ice-core records from Greenland and from the Antarctic to each other. And that, in turn, meant that the temperature records of the two polar regions could be accurately compared through time. More than 20 seesaw cycles were discovered in this way within the last ice-age cycle, with durations of one to several millennia.

In each cycle, cold conditions in Greenland coincided with increasing temperatures in Antarctica, in such a manner that the duration of cold Greenland conditions scaled with the amount of warming in

Antarctica. Next, Greenland experienced an abrupt warming (within decades), so that both peaked at roughly the same time. Antarctic temperatures then decreased while Greenland stayed relatively warm, followed by a rapid drop in Greenland temperatures that roughly coincided with the temperature minimum in Antarctica. Finally, as Greenland was in a cold phase, Antarctica was warming again, restarting the cycle (figure 17).

The seesaw's out-of-phase behavior between temperature changes in Greenland and Antarctica has been successfully explained in terms of heat redistribution within the oceans, and especially in the Atlantic Ocean. The mechanism involves the thermohaline deep-sea circulation component that is driven by NADW formation in the northernmost North Atlantic, as well as the surface circulation that compensates for this deep-water formation (see figure 7).

NADW is relatively saline owing to a subtropical surface-water influence that is carried northward by the Gulf Stream and North Atlantic Current. But if we take a closer look, then we see that this surface transport involves almost the entire surface of the world ocean. This is because the equator-crossing South Equatorial Current, which transports South Atlantic surface waters northward, feeds part of the Gulf Stream. In turn, the South Atlantic surface waters are fed by leakage from the warm Indian Ocean around the tip of South Africa, by means of the Agulhas Current—we refer to this as the "Agulhas leakage." But the connections don't stop there. The Indian surface waters that feed the Agulhas leakage derive from surface flows in the wider Indian Ocean. And the surface flows in the Indian Ocean are well connected with warm Pacific surface waters by means of powerful surface-water flows through the sea straits between Southeast Asia, the Indonesian islands, New Guinea, and Australia. This way, virtually the entire surface ocean is connected.

This worldwide surface flow forms the return transport that compensates for sinking of new deep waters in specific, focused regions, like the northernmost Atlantic. After sinking and spreading through the world ocean, the deep-water circulation causes a much less focused net upwelling and up-mixing through most of the world ocean. Eventually the former deep water is upwelled or up-mixed enough to become part of the surface circulation, which eventually reconnects with, and resupplies salt to, the regions of new deep-water formation.

The North Atlantic surface circulation, including the Gulf Stream and North Atlantic Current, is predominantly wind driven, by the subtropical trade winds and the midlatitude westerly winds (top panel in figure 6). The circulation crosses the North Atlantic from North Carolina in the United States toward Ireland. This landmass deflects most of the wind-driven component in a southward direction. But part of the distinctly warm and salty current system bends northward, toward the high-latitude NADW formation sites, and there it compensates for the transformation of surface water into deep water (NADW). The associated northward salt transport is very important for the formation of NADW. In this way, it seems like a classical chicken-and-egg problem: what comes first, the salt transport or the deep-water formation that pulls in the salt transport? But it should more appropriately be considered as a closely interconnected system, in which processes relate to each other as feedbacks; one thing leads to the other, and it makes little difference what you start with.

The system can be disrupted. Net freshwater input into the North Atlantic, for instance due to ice melt or to a strong reduction in evaporation, can temporarily decrease or shut down NADW formation and its associated northward transport of warm and salty water. Because this stops the northward transport of warm water through the Atlantic Ocean, it cools the northernmost Atlantic regions. At the same time, the heat that is not being pulled northward is instead retained in surface waters of the South Atlantic, from where the mighty Antarctic Circumpolar Current distributes it over the southern hemisphere. So the longer the north stays cold, the more heat builds up in the south—in other words, the longer the cooling in the north (as recorded in Greenland), the more the south warms up (as recorded in Antarctica).

As NADW formation is suddenly restarted, the north (Greenland) abruptly warms up, and warming stops in the south. This causes Antarctica's warming to top out, followed by a cooling trend. While NADW formation continues, Greenland stays warm. The south (Antarctica) gradually loses heat because of northward heat transport and therefore cools. When NADW formation becomes inhibited again,

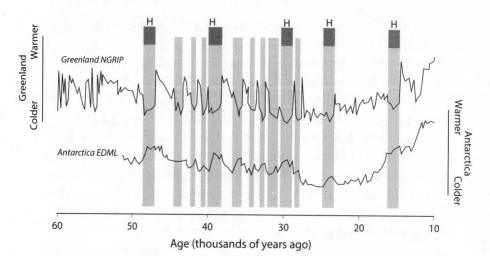

Figure 17. Schematic comparison of temperature reconstructions from ice cores for Greenland and Antarctica during the last ice-age cycle. The *light gray bars* indicate phases in the bipolar temperature seesaw during which Greenland is cold, and Antarctica is warming. *Dark gray blocks* with letter H indicate approximate intervals of Heinrich events of ice-rafted-debris deposition in the North Atlantic due to massive iceberg calving. EDML, EPICA Dronning-Maud Land, where EPICA stands for European Project for Ice Coring in Antarctica; NGRIP, North Greenland Ice Core Project.

the cycle is restarted: Greenland cools abruptly, and renewed heat retention in the south warms up Antarctica. This is the essence of the bipolar temperature-seesaw process, as first outlined in the early 1990s by Thomas Stocker, and by Tom Crowley (see figure 17).

In the North Atlantic region, including Greenland, the most striking feature of the seesaw cycles is an extremely rapid and strong warming at times when NADW formation was fired up again. Within a few decades, Greenland temperatures shot up by 8°C to 15°C, and European temperatures by 5°C or more. Temperatures then stayed high, albeit with a gradual reduction over several centuries, followed by an abrupt cooling event when NADW formation was reduced or stopped. The cooling events were only slightly less dramatic in size and rapidity than the warming events. These variations happened with an apparently regular spacing of about 1500 years, or multiples of 1500 years. Expressions of this variability have been recognized all over the northern hemisphere, affecting, for example, monsoon intensities and windblown dust variations, and including intermediate-water circulation and oxygenation in the northwestern Indian Ocean and in the North Pacific.

Seesaw events are poorly developed during glacial maxima, and not present in any significant form during interglacials. Overall, the characteristic seesaw variability seems to have been best developed at times when global sea levels stood between about 90 and 50 meters lower than today—in other words, clearly during times of "intermediate glaciation," given that sea levels stood some 125 meters or more below the present during full glacial conditions.

Ocean and climate models indicate that the deep-sea circulation in glacial oceans was more sensitive to change than that in interglacial oceans, which may explain why seesaw events are not found in interglacials. But it is not yet evident why seesaw events were less pronounced during glacial maxima. Part of this uncertainty comes from the fact that we still don't know what exactly caused the events. Concepts that have been proposed include: iceberg calving and melting, ice-dam breaks and associated rapid meltwater input into the ocean, changes in the salt content of North Atlantic surface waters due to net evaporation changes and/or changes in saltwater inflow around the tip of Africa (Agulhas current), sea-ice changes in the North Atlantic due to sea-level-related changes in shallow shelf areas, and variations in the moisture balance between the Atlantic and the Pacific. All of these can explain part of some of the events, but not all aspects of all of them. Some of the potential controls listed here were different during glacial maxima than at times of intermediate glaciation, and therein may lie a clue why seesaw events are not well expressed during glacial maxima. There is a lot of research activity directed to answering this question. It has become especially pertinent because recent observations suggest that NADW may be undergoing some degree of slowdown as we speak.

For about seven of the more extreme events during the last 100,000-year ice-age cycle, the North Atlantic contained remarkable layers of ice-rafted debris (IRD), or concentrated layers of sand and gravel-sized dropstones. These layers have been named Heinrich layers after the German researcher who first described them, and the climate events during their deposition are called Heinrich events (the five youngest of these episodes are indicated in figure 17). At these times, for several hundreds to perhaps a thousand years, the North Atlantic must have been filled with melting icebergs that had calved off the large ice sheets of the time. The melting ice also caused measurable (using delta-^{18}O) freshening of surface waters. Geochemical and mineral-magnetic measurements of NADW circulation indicate that NADW formation was almost completely inhibited at these times, most likely because of the strong fresh meltwater input. Each of the Heinrich events in the north corresponds to a phase of remarkable warming in the southern hemisphere by up to 3°C or 4°C, in line with the seesaw mechanism.

For the more than a dozen other, weaker seesaw events visible in ice-core and marine-sediment-core records, we don't really understand the causes of the critical salinity changes in the North Atlantic. It may have been a much more subtle freshwater forcing just in the locations of NADW formation, compared with Heinrich-event freshening over large parts of the North Atlantic.

We're also largely in the dark about what determined the apparent 1500-year time scale for the seesaw cycles, and why sometimes cycles are skipped or merged into durations of multiples of 1500 years. Several internal oscillations between components of the climate system have been suggested. Others invoke responses to some aspect of—less well-understood—solar variability. The jury very much remains out on this question.

The most recent North Atlantic freshening event with notable reduction of NADW formation took place about 8,200 years ago. It is known as the "8.2 ka event," where ka stands for kiloyears ago, or 1000 years ago. At that time, a final remaining ice dam in North America broke through, and a very large ice-dammed lake drained into the North Atlantic through the Hudson Strait. The resultant 150-year long 8.2 ka event saw a sharp cooling of several degrees centigrade in Greenland and on continents around the North Atlantic, and there seems to have been a negative impact on northern-hemisphere summer-monsoon intensity as well. This event was part of the final death throes of the last ice age. The deglaciation was complete at around 6000 years ago.

Since this event, there have been no more significant meltwater input pulses with widespread impacts on NADW formation. More minor climatic variations, such as the Little Ice Age, have still taken place since then. But these were mostly variations around the modern climate theme, and none seems to have involved significant changes in ocean circulation, properties, or life. Worryingly, however, initial signs of ocean change have reappeared over the last century or two.

*****

To recap, we have seen that the Cretaceous-Paleogene extinction event of 66 million years ago may have been disastrous to life, but that the general prevalence of greenhouse climate conditions was not over. These conditions continued until about 46 million years ago, when the first signs of intensifying cold appeared in both hemispheres. But it was not until 33.9 to 33.6 million years ago that the first large ice sheet became established on East Antarctica, in a distinct two-step process that beautifully illustrates the influences of astronomical control of climate under certain conditions (a specific CO_2 threshold). This ice mass has remained variably present since then. It was joined by a variable West Antarctic Ice Sheet by about 15 million years ago. The next big development was a beginning of glacial processes on the northern hemisphere from about 10 million years ago, with initial formation of a minor ice sheet on southern Greenland, which eventually expanded 3.3 million years ago.

Next, the first significant glaciation on the North American continent appeared some time around 2.5 million years ago. This brought us firmly into the reality of the ice ages, with early ice ages following a cycle of 41,000 years' duration, and later ice ages following a (sawtooth-shaped) cycle with an average 100,000-year duration. Along the way, we discussed how sea-level fluctuations are reconstructed, as well as the carbon-cycle implications of the ice-age cycles, highlighting the special role of the ocean in storing carbon during glacials, thus resulting in lowered atmospheric CO_2 concentrations. Finally, we looked at superimposed bipolar temperature-seesaw cycles with durations of a few thousand years. We saw that these were strongly related to heat-redistribution processes associated with changes in the intensity of NADW formation. And ultimately, by about 6000 years ago, all the big ocean changes were found to be over.

CHAPTER 8

FUTURE OCEANS AND CLIMATE

The ocean has been remarkably invariant over the past few millennia. But one or two centuries ago, several big new changes announced themselves—it's time that we take a look at those. In this chapter, human-caused ocean acidification and warming are evaluated from the perspective that we have gleaned from the geologic record. First, however, we need to take a look at the scale and rate of human-caused net carbon emissions in comparison with geologic events, since these emissions lie at the heart of both ocean acidification and warming.

OUR CARBON EMISSIONS

Humans have, since the beginning of the industrial revolution, caused net emissions of some 420 GtC. This has driven an extremely fast rise in atmospheric CO_2 levels by 120 ppm, from typical interglacial values of about 280 ppm before the industrial revolution, to around 400 ppm today (see figure 11). The current level is equivalent to that last seen about three million years ago, when global temperature was some 2°C to 3°C higher than today and sea levels reached 10 to 30 meters above the present level. Most emission projections that account for ongoing trends argue that we are likely to exceed levels of 700 ppm by the end of this century. Such levels were last seen in times before 34 million years ago, before the glaciation of Antarctica, when deep-sea temperatures were some 10°C or more higher than today.

Given the close long-term relationship between temperature and CO_2 levels throughout Earth's history, the current CO_2 increase may be expected to cause dramatic changes in climate and the oceans. We don't immediately have to worry that the Antarctic ice sheet will completely vanish in the short term, or that the deep-sea temperature will shoot up, as such changes take hundreds to thousands of years to reach completion. But Earth's history does teach us that the slow feedbacks—notably deep-ocean warming, global ice-volume changes, and global carbon-cycle adjustments—will continue to adjust to our disturbance over many centuries to millennia, even if we stop emissions. Climate changes move like massive freight trains: they will keep going once they are set in motion. The impacts of our emissions will not stop in the year 2100, but will continue to keep adding up as the slow feedbacks continue to adjust. Thus, climate will get "locked in" on a long-term trajectory to a much warmer greenhouse-type world with much less ice if we don't fight the CO_2 rise rapidly, before the slow feedbacks gain too much momentum. Once these responses get properly going, any real reduction in CO_2 levels (if and when we can engineer ways to create that) would take equally long to be registered by the slow feedbacks.

So our only bet is to try to slow the train down before it has come up to speed. As is, the slow-feedback train is visibly coming into motion. Ocean warming is mainly restricted to the surface 1000 meters or so, but initial signs of warming in parts of the deep sea have been reported as well. Mass loss from continental ice sheets is measurably accelerating, and the ice streams that feed this mass loss have become increasingly active. Deforestation has deeply perturbed the carbon cycle and the operation of ecosystems on land, and it will be a challenge to responsibly reverse that. The expansion of aridification over large areas of the world is similarly affecting land-based ecosystems. Overfishing and acidification are upsetting marine ecosystems. All these ecosystem changes are affecting the carbon cycle. Eventually, penetration of our carbon emissions into the deep ocean will cause further changes in the carbon cycle, as carbonate compensation plays out over thousands of years to come.

Even without further increase, the net emissions that we have effected already (420 GtC) represent a stupendous amount of carbon.

It is equivalent to a solid graphite column with a diameter of 25 meters, stretching all the way from Earth to the Moon. It is equivalent in size to 60% of the total carbon exchange between the ocean and the atmosphere-biosphere system associated with an ice-age cycle, except that it's worse. It's worse because some 90% of the human-caused emissions do not represent a shuttling of carbon between components of the climate system, as was the case for the glacial-interglacial carbon exchanges. Instead, it concerns a net input of external carbon: we have put *extra* carbon, once safely locked within the Earth, back into the climate system. So we should expect more long-term impacts until natural or human-made processes can remove it again.

But I haven't even touched upon the worst part yet. The worst aspect is the rapidity at which these emissions have taken place. The human-caused emissions were realized in as little as 200 years, or less, so the average human-caused rate of net emissions was 2.1 GtC per year. The actual rate of emissions today is five times that average value—it is close to 10 GtC per year. In contrast, glacial-interglacial transitions may have been fast by natural standards, but still took some 10,000 years. The average natural rate of change, therefore, was 0.07 GtC per year. Hence, average human-caused carbon emissions have been 30 times more rapid than the average natural rate of change during the ends of ice ages. Earlier, we calculated that the average rate of human-caused carbon emissions is about five times faster even than that of the dramatic PETM emissions. And the actual emission rate in the last couple of years was more like 30 times higher than that of the PETM.

We saw that the end-Permian extinction is the only event in Earth's history for which at least some evidence suggests external-carbon injection at rates anywhere close to those of the human-caused emissions. But this rate is debated and may have been considerably lower. The end-Permian event was also the time of the world's worst extinction event ever: 96% of marine life and 70% of land-based life became extinct. The end-Permian isn't a perfect analogy, because our influence on the climate is unprecedented. But, without doubt, the end-Permian event can tell us a lot about the potential depth of the modern crisis in a qualitative sense.

The end-Permian event was characterized by a triple whammy of ocean acidification, rapid warming, and ocean anoxia, as well as the widespread species extinctions. The combined scale, rapidity, and duration of these impacts made the end-Permian a severe mass extinction. The modern impacts and associated changes are at least as fast. They are not (yet) projected to reach the same scale, but the warning is clear: continuing emissions at current levels risks bringing us ever closer to an end-Permian-type situation.

This brings up an almost inevitable question: *Does this mean that a mass extinction event is coming?* Or, alternatively: *Has one possibly started already?* To assess that, we can compare how fast extinctions are taking place today with the natural background rates of extinction. Background rates are the rates of occasional random extinctions that continuously occur, and they amount to an average of up to 2 species per year for every 1 million species on Earth. A 2011 survey estimated that humans share Earth with 8.7 million species of life, give or take 1.3 million. Note, however, that such estimates are subject to enormous changes because it is very hard to work out, not only because of obvious difficulties in recognizing and counting all species of the more obvious animals, plants, fungi, algae, etc., but especially because the number of microbial species increases with every new study. If we assume that the number is 10 million, then the estimated background extinction rate would be 20 species per year.

There has been a lot of research on modern extinction rates and how these compare with the natural background rate. Recent results show that the world is currently experiencing an exceptionally high rate of extinctions, at 1000 to 10,000 times the natural background rate. Even just for vertebrates, the estimated modern extinction rate is 114 times higher than the vertebrate background rate. Some estimates of the overall current rate of extinction already reach more than 10 per day. This is why researchers consistently speak of the impending, or—more precisely—already ongoing, sixth major extinction event.

How can we compare this enhanced current rate of extinctions with those of major extinction events of the past? For this, it is crucial that we remember the depth of time that applies to the geologic past. When investigated up close, even the geologically almost instantaneous end-Permian mass extinction turns out to have stretched over tens of thousands of years. As always, everything is relative: extinctions seem instantaneous on the incredibly deep geologic time scales, but they took forever when viewed on our short human time scales. If we make a bold assumption that the number of species on Earth before the end-Permian extinction event was similar to that of today (say, about 10 million), and that about 90% of those were terminated over the 60,000-year duration of the event, then we have an estimated end-Permian extinction rate of 150 species per year. Remarkably, this is only 7.5 times the background extinction rate calculated above. From this perspective, even a mass extinction event like the end-Permian in full flow might actually be quite hard to spot on human time scales. And contrast this with the fact that current extinction rates are 1000 to 10,000 times the background rate (or still about 100 times if we consider only vertebrates). These rates completely dwarf our rough estimate for the end-Permian event; even if we got that estimate wrong by a factor of 10, then current rates of extinction are still higher! The importance of this cannot be overstated. We owe a huge debt to researchers for spotting that a massive mass extinction is trying to sneak up on us. Having done so, they have allowed us time to act.

The hard data of an exceptionally high modern extinction rate clearly reject the frequently voiced claim that what's happening "*is just natural variability.*" This claim seems to be motivated by a Hollywood-style expectation that a mass extinction should see the whole world dead and barren from one day to the next. If we instead accept that things occur more gradually, and that we are on the cusp of the sixth major extinction event, then should we be worried? Would it affect us humans? This is anybody's guess, but geologic data do offer some perspective.

First, we need to acknowledge that our impact on the environment is not limited to "just" that of our very fast and large external-carbon emissions, but instead also includes widespread pollution and eutrophication, as well as major physical ecosystem degradation through deforestation, large-scale monoculture farming, overfishing, river-water diversion, industrialization and construction in many sensitive environments, and so on. Such a simultaneous multipronged attack on all of the Earth's major environments (marine, freshwater, on land, and in the air) is truly unprecedented in Earth's history. The extremely high rate of extinctions of modern times shows that ecosystems everywhere are failing as a result—and, crucially, these ecosystem collapses are not "at some time in the future," but instead are systematically happening already. Hence, it really is not exaggerated or alarmist to talk about an impending major mass extinction, the beginnings of which may already be in progress. Given the deep compound nature and fast rate and amplitudes of our impacts on nature, it is then also valid to compare our potential pathway with the worst natural mass extinction that we know; the end-Permian event.

Second, we should remember that humans are not prokaryotes that can survive wildly different and extreme environments. Instead, we are complex eukaryotes that for our very existence rely on an intricate food web that consists of other complex eukaryotes. Therefore, if we do allow the world to slip away from us in a similar way that it went at the end of the Permian, then we more than likely won't come out of it very well, if at all. Earth would continue to exist, and life in a general sense would most likely endure, but the ecological "reset" of a mass extinction would in all probability remove almost everything bar the simpler and more extreme-adapted life-forms.

Of course, nobody can exactly predict this because of the many aspects of chance involved in the processes behind extinction, and because of the unpredictability of human ingenuity. My vote, however, is to err on the safe side, to heed the warnings of Mother Nature's past experiments.

Given that nature has dealt with carbon before, and at higher levels than we have in today's atmosphere, it is tempting to assume that nature will somehow take care of things for us. But it's a position of convenience; if we believe it, then we could just keep going the way we are going, expecting nature to come to the rescue. Unfortunately, that is not borne out by Earth's history.

For example, the exceptional Devonian-Carboniferous CO_2 reduction by almost 5000 ppm was brought about by the unique development of a completely new design of land plant that subsequently conquered the planet, and still took about 100 million years. This implies an average rate of change of -50 ppm per million years, or -0.00005 ppm per year. Although this was a CO_2 extraction event of great magnitude, its rate of CO_2 change still pales into insignificance relative to the rate at which we humans are pumping CO_2 into the atmosphere. We have caused a rise of about 120 ppm in about 150 to 200 years since the onset of the industrial revolution, which implies an average rate of rise of +0.6 to +0.8 ppm per year. And in recent years, the actual rate of rise has been higher than +2 ppm per year. These average human-induced CO_2 emissions are *12,000* to *16,000* times faster than the natural CO_2 extraction of the dramatic Devonian-Carboniferous episode. In recent years this ratio has gone up to *40,000* times.

So, if we can exclude significant assistance from carbon burial on land in cleaning up our emissions, then how about large-scale carbon burial in the oceans? In the well-studied, youngest Mediterranean sapropel (S1), on average 0.65 grams of carbon was extracted per square meter every year. For a very rich sapropel (S5), this value was up to four times higher, which gives 2.6 grams of carbon per square meter per year. Now let's assume that an oceanic anoxic event were to develop in the entire world ocean, with black-shale deposition everywhere below 1000 meters' depth, and that this black shale were to extract carbon at the same rate as a very rich Mediterranean sapropel. The surface area of the world ocean at 1000 meters' depth is 318,000 billion square meters. Thus, we calculate a total carbon extraction of almost 830,000 billion grams, or 0.83 GtC, per year. This represents the estimated annual carbon extraction by a devastating process in which the entire world ocean below 1000 meters' depth becomes anoxic, which means that almost everything dies. Even such a dramatic process could offset only about 40% of our average yearly carbon emissions of the last 200 years, or only 8% of the yearly carbon emissions of recent years.

Such comparisons put the immense scale of human-induced carbon emissions into a sobering natural perspective. Even considering extreme mechanisms, there is nothing in nature that can offset more than about one-tenth (1 GtC) of our annual emission rates of recent years. Even if nature's extraction mechanisms were to be fired up to full capacity, then 90% (9 GtC) of our yearly emissions would still need to be dealt with in different ways. These different ways would need to be rooted in human-driven emission control and carbon capture and storage.

If we compare our emissions with natural rates of carbon injection, we find that the fastest known event (at the end of the Permian period) might just about have achieved similar rates to the current emissions. But most events remained limited to a maximum of about 20% of the current rates, as in the case of the PETM. Both of these fast natural-injection events were associated with dramatic ocean acidification, because natural carbon extraction was too slow to deal with the carbon injections. Natural compensation for the added carbon in these events seems to have lasted at least 10 times longer than the time scale of the injections. This again indicates that natural processes can only extract carbon at rates that are no more than one-tenth of the rates of modern carbon emissions.

The above provides critical context to the often-voiced opinion that current climate change might be *"just a blip"* against larger natural variations. Nature's main mechanisms of response to an external-carbon injection—organic carbon burial, carbonate compensation, and weathering—may be extremely powerful in terms of the total CO_2 reductions that they can achieve, but they are also excruciatingly slow, both in real terms and in comparison with the very fast rate of human-induced CO_2 emissions.

Like a human supervolcano, our fossil-fuel injection of external carbon has created a large and fast-growing CO_2 spike in only 150 to 200 years. Based on the slow rates of the natural CO_2 extraction processes (above), as beautifully illustrated by the recovery from the PETM carbon injection, it will take tens to hundreds of thousands of years for nature to bring the modern carbon spike back under control, if we leave it to its own devices. So, yes, eventually nature can, and will, remove our carbon. Just don't wait up for it.

CONSEQUENCES

A direct consequence of our industrial-age net emissions of external carbon is ocean acidification. The emissions are so fast that they overwhelm the natural responses that might offset the resultant acidification. Before carbonate compensation can even begin to slowly compensate for the CO_2 increase, the carbon needs to penetrate into the deep sea, and that takes many centuries to even a millennium. In all, carbonate compensation works over typical time scales of 5000 to 10,000 years, and it will take several times that typical time scale to negate the total CO_2 injection. Owing to the unbalance between the fast addition of carbon and the slow and delayed responses to it, the acidity of the ocean waters increases rapidly.

Ocean pH has dropped by 0.1 units already, from a preindustrial value of 8.2 to 8.1 today, which represents a 25% increase in acidity over the past two centuries. With continuation of today's emissions, an ocean pH of about 7.8 or 7.7 is expected by the end of this century. The inferred pH drop of 0.3 to 0.4 units is well beyond what fish and other marine organisms can tolerate in the laboratory without very serious implications for health, reproduction, and mobility. Most organisms experience severe problems with a pH drop of only 0.2 units. And the extreme rapidity of the change prevents meaningful adaptations to the increasing acidity.

In short, we can expect very serious consequences for the marine ecosystem. Carbonate producers such as corals will be especially affected. The World Wildlife Fund *Living Blue Planet Report* of 2015 infers that the cumulative impacts of acidification, pollution, eutrophication, and physical damage may drive coral reefs to extinction by 2050. Coral reefs are critical nursery grounds for many oceanic animal species. Losing them will have a very profound impact throughout the oceans, especially when considered alongside other major pressures such as overfishing and pollution. Moreover, it is becoming increasingly evident that the double whammy of temperature rise and acidification will fundamentally reduce diversity and numbers of key species throughout the marine ecosystem. This could induce a species collapse from the top of the food web downward.

Until recently, ocean acidification has remained below most people's radar. But it is developing fast enough, and to a sufficient scale, to unleash devastating impacts on the global oceanic ecosystem in the not too distant future. These will affect a very broad range of food resources for humanity, which are particularly important in many of the world's less affluent nations. Ocean acidification therefore needs to become an integral part, on a par with warming, of the necessity for solutions to our fossil-fuel addiction.

Because ocean acidification is a direct consequence of our net carbon emissions, it will not be remedied at all by geoengineering projects that focus only on reducing the radiative forcing of climate and not the actual carbon emissions. This includes proposed concepts to increase cloud cover by sea-spray injection or other means, or to reduce solar irradiation by placing arrays of mirrors in space, or similar. To avoid ocean acidification and climate warming alike, there is no alternative to reducing our net carbon emissions. This then places emphasis on geoengineering projects that aim to remove carbon from the atmosphere.

Projects to remove carbon from the atmosphere are developing fast, but none has made it into industrial-scale testing yet. One of the potentially more feasible approaches has long been advocated by the Dutch geochemist Olaf Schuiling. It focuses on mining, grinding up, and widely distributing rocks that are rich in chemically easily weathered silicate minerals, such as olivine, on land and in coastal regions. Especially in warm and humid climates, weathering will be intensive, consuming CO_2 from the atmosphere. The process of making these minerals available in artificially increased abundances accelerates atmospheric CO_2 reduction due to weathering, and the end products are completely natural. This is an interesting concept, and one that according to calculations has the potential to make a difference.

But there are some complications. First, the mining, grinding, and distribution of the rocks is costly, and it also has to be done in a carbon-neutral manner to avoid first emitting extra carbon for the sake of removing it later. Second, we know that the oceans have taken up about a third of our total carbon emissions to date, in a bid to offset the rise in atmospheric concentrations. This partitioning works the same way for a negative emission (that is, a reduction) of atmospheric CO_2. Because the ocean will thus feed CO_2 back into the atmosphere as we start reducing atmospheric CO_2 levels, we will need to remove much more CO_2 than we would think from looking at the atmospheric values only; just under twice that amount, actually. Hence, we will need a gargantuan amount of olivine. If one calculates the amount of dunite (a rock that contains more than 90% olivine) that would need to be mined for a 50 ppm atmospheric CO_2 reduction, then it equates to a mass of about 200 billion cubic meters. This is roughly equivalent to creating a hole about 170 times larger than the deepest—and almost largest—man-made hole on Earth: the

Bingham Canyon Mine or Kennecott Copper Mine in Utah. This is a colossal task, especially if it's to be done in a carbon-neutral manner. But it's of the sort of practical challenge that humans have proven themselves to be good at resolving—there might be a way.

Other geoengineering approaches for lowering atmospheric CO_2 levels are also far from an implementation stage. Of these, artificial ocean fertilization has received the most attention. But it remains unsure whether the carbon drawn into the ocean by this process would remain there, or whether most of it would be vented out again, and on what time scales this would happen. As we have seen, even under highly productive natural conditions, only a fraction of the total surface organic carbon production is eventually buried in sediments, while the majority is respired into the water column. From there, it participates in the active carbon exchanges between oceans and atmosphere. So ocean fertilization may result in a temporary reprieve only. Arguably, it is only over the very long time scales of carbonate compensation that this mechanism would result in a significant long-term net carbon reduction. But then again, maybe a temporary reprieve is all we need, giving us time to spin up alternative and more permanent carbon-removal mechanisms.

The best-known consequence of human-caused net carbon emissions of the industrial age is global warming. Earth's history reveals a sound relationship between CO_2 levels and global temperature (and glaciation state), even though the relationship is complicated and regularly obscured by myriad superimposed sources of variability. The relationship is predictable on the basis of what we know about the radiative greenhouse properties of CO_2. Complicating influences mainly arise from solar-output change, concentrations of other greenhouse gases (especially CH_4), and powerful radiative feedbacks such as those involving ice-albedo, vegetation- and land-surface-albedo, and aerosol impacts. Other factors that can complicate the relationship between CO_2 and temperature are less immediately obvious.

Besides radiative forcing and feedbacks, Earth's temperature also depends on very slowly changing boundary conditions, where particularly important ones include the layout of the continents and mountain ranges, and the evolution of new vegetation types through time.

For example, the land-sea distribution over the planet is important because seawater is less reflective than land surface, so that the globe was more reflective at times when the continents were concentrated in low latitudes. Also, it has recently been suggested that about half of the deep-sea cooling from the Middle Mesozoic warm conditions to the present may have resulted from plate-tectonic influences on the distribution of continents and ocean passages. Evolution of new types of vegetation cover changes the overall reflectivity of the planet through time, since—for instance—vegetation is less reflective than bare rock, lush forests less reflective than shrubland, and wooded grassland more reflective than woods with fern-based undergrowth. Mountain ranges at high latitudes can promote the formation of ice caps, as was suggested for the early development of the Antarctic ice sheet by merging of individual ice caps. Major north–south mountain ranges like the Rockies or Andes can cause virtually permanent swings in the jet stream that cause drought in some zones, and that promote moisture transport to high latitudes in other zones, fuelling snow and ice formation. Mountain chains can also inhibit atmospheric freshwater exchange between ocean basins, and thus affect deep-ocean circulation patterns. And so on. The ever-changing nature of the boundary conditions has an important implication; namely, that no time in Earth's history has ever been identical to any other time in Earth's history. In that sense, each period of time was truly unique in its own right, including the present. For me, this is one of the aspects that keep past climate research so interesting—there's always something new to learn from every time interval studied.

Yet through it all, and despite superimposed divergences, one general rule of thumb has always applied: the world was warm in times of high CO_2 levels, and cold in times of low CO_2 levels (see figures 11 and 14). Without the greenhouse influences, these warm and cold variations cannot be explained in terms of the Earth's energy balance. And CO_2 has been the main greenhouse gas to be considered ever since the Great Oxygenation Event.

Another consistent observation is that high-CO_2 greenhouse periods of the past were characterized by weaker equator-to-pole temperature gradients than today. In technical terms, we use the term polar amplification to describe the enhanced sensitivity of polar climates to changes in the global climate state. In a world with high latitudes dominated by tundra, bare rock, and ice and snow, the poles become disproportionately cold because of land-surface and ice-albedo feedbacks. In times that were globally warm enough to have no ice and snow, this polar cooling influence did not exist. Instead, high latitudes contained forests that are less reflective than tundra and bare rock, which promoted polar warming. In addition, warmer times had more water vapor in the atmosphere, which amplifies greenhouse warming on a global scale. Modeling studies in addition suggest potentially important effects of interactions between ocean heat transport and middle- to high-latitude convective cloud cover, and of interactions between high CO_2 and CH_4 levels and the formation of polar stratospheric clouds. The latter might cause some 7°C excess polar warming by preventing heat loss during the long polar night.

Today, we are losing polar ice and snow at an alarming rate. Vegetation—including forests—invades the high latitudes. Given the many feedback processes that are known to affect polar regions, we may expect future warming and ecosystem pressures to be much more pronounced in polar regions than at low latitudes.

The rapid warming due to human-caused net carbon emissions has direct effects on the habitat ranges of both plants and animals. On land, species' habitat ranges are shifting upward with warming, so that species adapted to high mountain environments are being squeezed out. Both on land and in the oceans, species' habitat ranges are noticeably shifting poleward. In consequence, records at single locations are showing changing species compositions, with more and more warmer-adapted species moving in and colder-adapted species moving away (in a poleward direction). This changes the competitive environment, and hence the very fabric of the food web. In addition, it increases the chances of foreign-disease introduction.

Sessile marine species are fixed to the seafloor and cannot easily migrate. They are therefore especially susceptible to the adverse effects of increasing temperatures. A well-known consequence is coral bleaching, which occurs when corals expel their photosynthetic symbionts when under stress, most notably by warming. These corals become more susceptible to diseases and mass mortality. Together with coastal eutrophication, increasing temperatures also drive algal proliferation, another big contributor to mass mortality in reef environments.

The myriad impacts of warming, added to those of acidification, highlight that humanity's extreme net carbon emissions have us heading for seriously challenging times with major, unprecedentedly rapid changes in both the marine and land-based ecosystems. In the oceans, expect to see changes in species' habitats, ecosystem functioning and food-web perturbations, large-scale mortality as well as invasion of foreign species (and diseases), regional oxygen depletion, and possibly even sharp events or cycles of energy redistribution within the ocean-atmosphere system. Some such effects are visible already, and have been documented in numerous detailed studies and assessment reports. Others develop more slowly. In any case, ongoing and increasing emissions are only going to intensify the impacts. And, as we saw before, nature by itself cannot fix things on time scales that are useful to us—it simply does not have the mechanisms to do so. We have to assist it.

Nature is full of complex responses and feedbacks. As a result, changes may be dramatic in some regions, and comparatively moderate in other regions. Similarly, some regions may become seriously impoverished, while other regions may even experience a net benefit because of newly emerging opportunities. This is why continuous monitoring and a flexible approach to responses are essential for timely adaptation and mitigation. Embracing change and the associated opportunities may minimize economical impacts, especially in weaker economies. But this requires that we accept reports of ongoing and impending change, and stop delaying essential action through endless debate, negotiation, and senseless attempts at sowing doubt against a veritable tide of evidence. In regions where impacts turn out to be profound, which can be identified early through careful monitoring, doggedly sticking to traditional ways will only ensure a painful crawl toward a sticky end.

A further impact of global warming of the ocean surface is that it will intensify freshwater cycling through the atmosphere, owing to stronger evaporation from the warmer oceans, and a greater atmospheric moisture capacity under warmer conditions. Cycles of evaporation and condensation involve a lot of energy, technically known as latent heat, and movement of more moisture through the atmosphere means that more energy will cycle through the atmosphere. Debate rages over whether we will see a stormier atmosphere in general, or perhaps fewer but bigger storms. Meanwhile, poleward expansion of the zone of tropical conditions leads to expansion of the areas affected by hurricanes (as they are known in the Atlantic) or typhoons (as they are known in the Indo-Pacific). This means that areas that were previously just outside the hurricane zones will become increasingly affected by hurricanes, with all of the associated consequences.

Combined with global sea-level rise, a general increase in storm intensities spells doom for exposed coastal regions. The zone between modern sea level and 10 meters above modern sea level represents just 2% of the Earth's total land area, but it is home to almost one-tenth of the global population. It also generates roughly a tenth of the gross world product of about 77,000 billion, or 7.7 trillion, US dollars. Frequencies of flooding extremes will increase dramatically even if sea level rises only 20 centimeters. For 80 centimeters' rise, frequencies of flooding extremes may increase by 1000 times or more along the east coast of Australia. This means that flooding intensities previously seen only once per century become very common indeed, occurring 10 or more times per year. Effectively, this implies that the place is permanently flooded or flood damaged. The outlook is roughly similar around most coastlines of the world. Coastal cities of the world won't be able to live with this, except through commissioning massive engineering projects to keep the sea out, or to move critical infrastructure further inland. Both require decisiveness, risk acknowledgement, and multipartisan action from governments. Except for certain enlightened—and wealthy—corners of the world, I'm not holding my breath that this will happen any time soon. In any case, many parts of the world will encounter significant financial impediments to taking the appropriate action.

Unfortunately, only little can be learned about storm intensity through Earth's history, as we have no reliable means to measure storms in the past. I will therefore not dwell on storms any further here, no matter how important they will be to humanity's experience of climate change.

The consequences of human-caused net carbon emissions for ocean circulation are less predictable. The geologic record unfortunately does not provide any hard and fast rule about deep-ocean circulation being stronger or weaker in cold or warm climates. But we expect particularly the impacts of warming and fresh meltwater addition into surface waters. During an interval of fast warming with enhanced addition of ice and snow melt to the ocean—especially at high latitudes—it becomes more difficult for surface waters to reach the high densities of waters that sank and spread into the deep sea before the warming and melt addition. In other words, it becomes harder to form sufficiently dense waters to displace existing deep waters. We may therefore expect at least temporary deep-water circulation problems during phases of strong and rapid warming. There is a potential positive-feedback effect, where reduction of deep-water formation and warming of surface waters negatively affect the efficiency of CO_2 uptake and thus its removal into the deep ocean, which then drives toward increasing atmospheric CO_2 levels.

So, should we fear a major deep-circulation disruption, possibly giving rise to a "seesaw" event in which regions in and around the North Atlantic experience rapid cooling, and the southern hemisphere a more gradual warming? It certainly cannot be excluded, especially if a large meltwater pulse comes from Greenland straight into the regions of NADW formation. And new insights into the mass loss from continental ice sheets, along with unprecedented observations of large-scale meltwater penetration into the interior of the Greenland ice sheet, suggest that we don't yet fully understand all of the surprises that may lie in store where ice-sheet reduction is concerned.

Although there is evidence that deep circulation may be more resistant to perturbations during interglacials than during glacials, the 8.2 ka event illustrates that very large and abrupt freshwater forcing may still have notable impacts. Estimates for the 8.2 ka flood volume vary widely, from values equivalent to about 4% of the Greenland ice mass up to (unlikely) maximum estimates of 15%. These floods were released geologically instantaneously, within a few years.

The geologic evidence indicates that we may expect weakening of NADW formation in relation to climate change and ice-sheet reduction, and recent observations suggest that such weakening may be developing already. As yet, however, it seems unlikely that a full-scale collapse will occur, with an associated major temperature-seesaw event, unless a massive meltwater pulse were to somehow escape from Greenland. If a catastrophic flood of 8.2 ka–event scale were to come from Greenland in the future, then all bets are off with regard to a potential future seesaw event. But even during the 8.2 ka event, NADW formation seems to have been considerably weakened but not stopped, leading to cold conditions in the wider North Atlantic region that were striking, yet rather regionally focused and mostly limited to several degrees centigrade, sustained over about a century. Any warming in the southern hemisphere remained barely noticeable, so there was no strong seesaw response that would give the event a global character.

In consequence, we may not need to worry too much about the potential for a large global seesaw event in the future, but we are likely to see continued weakening of NADW formation due to increasing meltwater inflow into the North Atlantic. This will be associated with anomalous sea-surface temperatures in the northwest Atlantic, as less warm subtropical water is drawn northward. Observations have highlighted the development of such a "cold patch" in the northwest Atlantic, and this may be a contributor to changes noted in the atmospheric conditions over the northeastern United States, with an increased potential for deep winter snowstorms. Ironically, this would be a case of increased wintery conditions due to a process related to global warming. It doesn't surprise any climate scientist, all of whom expect increasing extremes, but it may confuse some of the general public, as well as certain politicians in Washington, DC.

Having said that we likely won't need to worry too much about a large global seesaw event, I hasten to add that I would instantly revise my opinion if convincing signs were to develop that the increasing mass loss from the Greenland ice sheet foretells an impending catastrophic collapse of the ice sheet. Indeed, quite a few of my colleagues are more deeply concerned about an NADW collapse than I am. We should be closely monitoring developments to make sure that we remain prepared for action.

Sea-level rise is one of the most devastating, and virtually impossible to reverse, impacts of our net carbon emissions. It has been a key component of climate change throughout Earth's history, and inevitably will be so in the future. As we have seen, there are several processes that affect sea-level rise. In icehouse worlds like that of today, continental ice-volume change is the dominant control on time scales of several centuries to many tens of thousands of years. Yet, as far as ice is concerned over the past few decades, the most conspicuous change in the high latitudes has not involved continental ice volume. Instead, it has been the fast reduction of Arctic sea-ice cover. Since sea ice floats, its disappearance does not directly affect sea level—you may verify this by putting an ice cube in a glass of water and marking the water level, waiting for the ice to melt, and checking the level again. But the sea-ice reduction still worries sea-level researchers because it is intimately connected with surface-ocean warming, which is also a dominant process in increasing ice-sheet mass loss.

Sea ice, especially with a fresh snow cover, is very bright and reflective; it reflects up to 95% of the incoming radiation. When it disappears, open water is exposed, which at high latitudes reflects some 60% to 70% of the incoming radiation (reflection is much higher at the poles than at low latitudes because of the angle at which the sunlight strikes the surface). Sea-ice reduction therefore results in an increase in the absorption of incoming radiation, and thus in surface-ocean warming, which causes further sea-ice melting in summer and delayed refreezing in winter. These feedbacks cause initial sea-ice reduction to be amplified, and the sea-ice pack is steadily reduced. Summer sea-ice cover in the Arctic has now reduced so dramatically (30% loss in just the last 40 years) that shipping through the Arctic passages has become a viable prospect. All the while, more and more heat has been collecting in the high-latitude ocean surface.

Warming of high-latitude regions, and especially of surface waters, causes retreat as well as intensified streaming and calving of ice from continental ice sheets, especially when these terminate in the sea. Once ice or meltwater flows transfer from the continents into the ocean, they cause sea-level rise. Modern Greenland shows intensifying mass loss associated with these processes, and researchers find similarly worrying trends in West Antarctica. Together, these ice sheets easily account for more than 10 meters of potential sea-level rise. Even a sea-level rise of only 2 meters may displace almost 2.5%

of the global population, and it would have enormous economic impacts. The trillion-dollar question, therefore, is: *How fast will sea-level rise become in the future?*

For their future scenarios without additional emission constraints (representative climate pathways RCP6.0 and RCP8.5), the fifth assessment report of the Intergovernmental Panel on Climate Change (IPCC), *Climate Change 2014*, estimates that sea level by the year 2100 will reach—within uncertainties—between 0.3 and 0.8 meters higher than at the beginning of this century. Note that, currently, our emissions are progressing along the lines of the worst-case scenario RCP8.5, so the higher value is more relevant. A recent assessment of estimates by international climate and sea-level experts returned a wider range, between 0.3 and 1.3 meters. Studies that use past statistical relationships between sea level and other climate parameters, such as temperature or radiative forcing, project to values of 1 to even 2 meters' sea-level rise by 2100.

In yet another approach, my team and I used the rates of rise and response times of about 120 events of sea-level rise from the past 500,000 years. All of these events date to geologically recent times before humans became important on the planet, and therefore concern variability in a climate system that was quite similar to that of today, but which was completely natural (without human impacts). As a result, the values that were found illustrate the documented natural behavior in the current climate system; there can be no doubt that nature possesses the mechanisms to achieve the documented changes, and that it knows how to activate them. Using that geologic context, and comparing its projections with sea-level observations since the year 1700, we arrived at plausible estimates of up to 0.9 meters for 2100, and 2.7 meters for 2200, relative to the level of the year 2000.

Despite the disagreements and debate over exact values, the various projections robustly highlight that sea level will continue to rise, and will rise ever faster now that the ice sheets have started to respond to climate change. Between 1700 and 2000, sea level rose by only 0.3 meters. By 2100, it will likely have increased by another 0.5 to 1 meters. By 2200, some 2 meters may have been added again. The accelerating rate of rise has everything to do with how ice sheets respond—slowly at first, then speeding up—and with the increasing warming that results from feedback processes to sea-ice and ice-sheet reductions.

The numbers suggested here apply only if no unexpected, dramatic ice-sheet collapses occur. That is, the numbers assume that the ice sheets decrease only through regular melting and calving. There is disagreement about the likelihood of catastrophic ice-sheet collapses, and about the "hidden" physics that might be accelerated into a collapse. Some suggestions based on such more dramatic scenarios offer values of several meters of sea-level rise by 2100.

The really tough question is about what we can do. One train of thought draws on concepts developed by Tim Reeder and colleagues in the UK Environment Agency's protection concepts for London. It revolves around a coping strategy, in which careful plans are drawn up for an intermediate value of, say, 0.5 meters by 2100. At the same time, there would be an extension option to a plausible upper limit of about 1 meter, and built-in creation of awareness of a dramatic (if less likely) maximum of about 2 meters. Implementation of the protection then should go hand in hand with close monitoring of sea-level rise. If and when observations indicate that sea level is likely to exceed the first stage of implementation based on the intermediate scenario, then the plan's extension options are activated. Meanwhile, the potential maximum scenario of 2 meters is used to ensure that emergency plans and training are developed for activation without delay if the unexpected happens.

All of this may sound easy on paper, but the costs and infrastructure needs will be enormous. International discussion and agreement will therefore be needed, and poorer nations will require well-planned assistance. More fundamentally, a plan along such lines would first require international acceptance of the overwhelmingly held scientific opinion that substantial sea-level rise has become inevitable. Hopefully, the Conference of the Parties (COP)21 agreement about global warming that was reached in Paris in 2015 will prove to be the foundation stone for such acceptance.

The reason why I stated that substantial sea-level rise seems inevitable is because even draconian emission control won't make much difference in the short run as far as sea level is concerned. It remains important to ensure that we don't increase atmospheric CO_2 levels any further to minimize the other consequences that we have discussed. But it won't make much difference to sea-level rise because ice sheets take many centuries to start responding to warming. The study mentioned above, in which we investigated the geologic context of future sea-level rise, suggests that today's rather slow increase in sea-level rise agrees with expectations based on the typical time scales for such responses in the past half million years. Now that the rise is under way, the rate of rise is expected to keep accelerating over centuries as the ice sheets try to "catch up" with the unnaturally fast and large human-made radiative forcing of climate. This is why the rise over the remainder of this century will be much larger than the preceding rise.

Where will it end? At 400 ppm, current atmospheric CO_2 levels already are as high as they were around three million years ago, and they are rising fast. By 2100, the mean estimates of IPCC scenarios RCP6.0 and RCP8.5 are close to 600 and 1000 ppm, respectively. Current trends are closest to RCP 8.5, and you may recall that CO_2 values above 750 ppm have not been seen since 34 million years ago, during times when even Antarctica was free of ice. Let's assume that emissions will be controlled in such a way that we fix the atmospheric values between about 400 and 600 ppm. In that case, the geologic relationship between sea level and climate forcing over the past 35 million years suggests that we are committed to 10 to 35 meters of sea-level rise over a time scale of many centuries to millennia, as the full range of slow and inexorable responses and feedbacks that has now been activated plays out.

These values are not based on conjecture or assumptions, but on detailed geologic observations and process studies of the relationships between CO_2 levels, climate change, and global ice volume and sea level. We have also seen from the geologic data that nature has no mechanisms to get rid of our CO_2 on a sufficiently short time scale—either we help it by cutting emissions and removing carbon from the atmosphere, or we accept living in a high-CO_2 world for tens to hundreds of thousands of years. If we allow a high-CO_2 world to establish itself, then it will within decades to centuries be much warmer, ice volume will increasingly rapidly reduce, and sea level will keep on rising and rising. The choice is ours.

Incidentally, looking beyond 2100 is often considered irrelevant, given that electoral time scales operate over only a few years, and individual development projects over several decades. However, such a longer-term perspective is highly relevant to major infrastructure developments, such as overall city planning. Throughout Europe and Asia, the foundations of most city infrastructure date back centuries to millennia, as do most of the supporting agricultural and fishery traditions and transport routes. Even the younger developments in the Americas, Africa, and Australia have roots that date back hundreds of years. Taking this into account, it is clear that we do have to think beyond the current century when considering climate change and its impacts on civilization.

To limit the impacts of ice-sheet responses and thus sea-level rise, imminent CO_2 reduction is needed to below 350 ppm. To even hope to achieve this goal, we need to change our ways immediately and start extracting a lot more carbon than we emit. We have had our wild party, emitted the carbon, and enjoyed the benefits. Now it's time to clean up our mess.

EPILOGUE

Our journey through the history of the oceans has illustrated their unbreakable association with climate: whenever a major change happens in the oceans, climate is involved, and vice versa. We have also seen that the majority of the most dramatic episodes of change involved notable greenhouse-gas changes. Critical changes in the climate and ocean system involved some or all of the following: acidification, temperature change, major ice-volume change, net carbon injections or extractions, eutrophication and ocean oxygenation, and large-scale ecosystem changes.

This list must sound familiar. After all, myriad assessment reports and studies have—at times very publicly—emphasized that humanity's impacts on the Earth system deeply affect each and every one of these elements. Moreover, we're pushing the system faster than it has ever been pushed, with the only possible exception of the end-Permian event of 252 million years ago, which gave us Earth's worst mass extinction. To top it all, we are putting massive additional pressure on both the marine and terrestrial ecosystems through large-scale pollution, overfishing, deforestation, and monoculture development. These are unnatural pressures without equivalents in Earth's history, and they are caused by the actions of one single species. If our extremely fast push on all the natural components of change alone weren't enough to create a precarious situation, then surely these added unnatural pressures will send us to the edge of the cliff.

Mother Nature's patience seems to be up. There are clear warning signals of an impending sixth mass extinction; it may even have started already. Common sense suggests a prudent response of large-scale

action toward a responsible, sustainable future, and it is clear that
we're going to have to step up and help nature with removing the
carbon we have released. An accessible way toward this is through a
combination of large-scale reforestation, and reduction of emissions;
that is, not just limiting emission growth per year, but actually decreasing emissions year on year. If implemented imminently, such
action might still allow us to offset our emission spike on a time
scale of centuries to millennia. In contrast, the opportunity vanishes
if we dally even a decade or two before we start, hoping that carbon-
extraction technology will come to the rescue like a superhero. For
such superhero technology to be ready for large-scale implementation in good time (that is, immediately), we at least ought to have
gone through extensive industrial-sized trials already, and we haven't.
So the only sensible way forward is to start reforestation and emissions reduction, and meanwhile work hard to develop the superhero
technology that will be needed to finish the job.

I am often struck by how easily decision-makers seem to ignore
that the benefits of such remedial actions extend beyond climate and
the oceans, to many other levels. For example, emission reduction
will benefit population health, given that the number of deaths per
year caused by air pollution is estimated to double to 6.6 million
by 2050 if worldwide emissions continue unabated. Further benefits
include energy self-sufficiency and long-term cost-effectiveness, new
technological development and implementation that can support
the economy, and the establishment of a long-term sustainable society for future generations. Finally, there appears to be a strong causal
relationship between climatic events and human conflict across all
major regions of the world. Hence, measures to limit climate change
form an important step toward reducing the risk of major conflicts
and their associated range of challenges, such as humanitarian crises,
mass migrations of climate refugees, societal unrest, and military ex-
penditure. Overall, several major economical assessments have found
that acting now to curb climate change is economically much more
favorable than remaining inactive.

All we need is a forward look and the courage to step out of our
fossil-fuel comfort zone. Meanwhile, we should not forget to act on
the more immediately resolvable issues of pollution, eutrophication,

and food-web disturbances. It is time to escape the trappings of incessant rhetoric, arguments, deliberate sowing of doubt and confusion, and trying to shoot messengers, which serve only greed, self-enrichment, and ignorance.

All the signs are that time is running out fast. Earth's history shows us clear examples of what happens when Mother Nature gets really cross, as happened 252 million years ago. Or even when she just gets seriously cheesed off, as happened about half a dozen other times. We would be wise not to play dare with her against such stakes.

ACKNOWLEDGMENTS

Special thanks go to the colleagues and friends who offered the time and advice that were essential to improving the manuscript through its many versions. For fear of forgetting someone, I will not attempt to list all who helped shape the general line of ideas behind this book. But Andrew, Appy, Caroline, Ellen, Eric, Frank, Gavin, Gianluca, Ivan, Jim, Jimin, John, Juan, Katharine, Mark, Mathis, Penny, Phil, and Tom deserve a specific mention for specific discussions, explanations, edits, and suggestions to parts or all of the manuscript. Although only a handful of our discipline's pioneering researchers have been mentioned by name, I am equally indebted to all other colleagues. Our community's research is a collective endeavor in which new knowledge is being added all the time, no matter how small or large the discoveries. Together, we are unraveling the history and processes of change in Earth's oceans and climate, and learning about potential future directions—what could be more important? I specifically thank two anonymous colleagues who officially reviewed the text and made many excellent suggestions—your comments and suggestions have been very welcome. This book would not have come to be without long evenings in my office and a couple of excellent writing retreats; I thank family and friends for allowing me the time. Finally, I thank the Australian National University and the Australian Research Council for hosting me on an Australian Laureate Fellowship, which allowed the focus and time required for completing the project.

BIBLIOGRAPHY

GENERAL REFERENCE AND BACKGROUND

Alley, R. B. *The Two-Mile Time Machine: Ice Cores, Abrupt Climate Change, and Our Future*. Princeton, NJ: Princeton University Press, 2000.

Beerling, D. *The Emerald Planet: How Plants Changed Earth's History*. Oxford: Oxford University Press, 2007.

Benn, D. I., and D.J.A. Evans. *Glaciers and Glaciation*. 2nd ed. London: Hodder Education, 2010.

Benton, M. J. *Vertebrate Palaeontology*. 3rd ed. Oxford: Blackwell. 2004.

Briggs, D.E.G., D. H. Erwin, and F. J. Collier. *Fossils of the Burgess Shale*. Washington, DC: Smithsonian Institution, 1995.

Broecker, W. S., and T. H. Peng. *Tracers in the Sea*. New York: Lamont-Doherty Geological Observatory, Columbia University, 1982. http://eps.mcgill.ca/~egalbrai/Earth_System_Dynamics/Tracers _in_the_Sea.html.

Budyko, M. I. *Climatic Changes*. Washington, DC: American Geophysical Union, 1977.

Conway Morris, S. *The Crucible of Creation: The Burgess Shale and the Rise of Animals*. Oxford: Oxford University Press, 1998.

Cowen, R. *History of Life*. 4th ed. Malden, MA: Blackwell Scientific, 2005.

Dawkins, R. *The Ancestor's Tale*. London: Phoenix, 2004.

Denny, M. *How the Ocean Works: An Introduction to Oceanography*. Princeton, NJ: Princeton University Press, 2008.

Giddens, A. *The Politics of Climate Change*. Cambridge, UK: Polity, 2009.

Huddart, D., and T. Stott. *Earth Environments: Past, Present, and Future*. Chichester, UK: Wiley-Blackwell, 2010.

Imbrie, J., and K. P. Imbrie. *Ice Ages: Solving the Mystery*. Cambridge, MA: Harvard University Press, 1979.

Intergovernmental Panel on Climate Change (IPCC). *Climate Change 2014: Synthesis Report. Contribution of Working Groups I, II and III to the Fifth Assessment Report of the Intergovernmental Panel on Climate Change.* Core Writing Team, R. K. Pachauri and L. A. Meyer (eds.). Geneva, Switzerland: IPCC, 2014.

Kemp, T. S. *Fossils and Evolution*. Oxford: Oxford University Press, 1999.

Mackenzie, F. T. *Our Changing Planet: An Introduction to Earth System Science and Global Environmental Change*. 2nd ed. Upper Saddle River, NJ: Prentice-Hall, 1998.

Rahmstorf, S., and K. Richardson. *Our Threatened Oceans*. London: Haus, 2008.

Schultz, H. D., and M. Zabel. *Marine Geochemistry*. Berlin: Springer Science & Business Media, 2013.

Segar, D. A. *Introduction to Ocean Sciences*. 2nd ed. New York: W. W. Norton, 2007.

Turcotte, D. L., and G. Schubert. *Geodynamics*. 2nd ed. Cambridge, UK: Cambridge University Press, 2002.

Wicander, R., and J. S. Monroe. *Historical Geology: Evolution of Earth and Life Through Time*. 3rd ed. Belmont, CA: Brooks/Cole, 2000.

Williams, R. G., and M. J. Follows. *Ocean Dynamics and the Carbon Cycle: Principles and Mechanisms*. Cambridge, UK: Cambridge University Press, 2011.

Woodward, J. *The Ice Age: A Very Short Introduction*. New York: Oxford University Press, 2014.

KEY SOURCES FOR CHAPTER 1

Chen, X., and K.-K. Tung. "Varying Planetary Heat Sink Led to Global-Warming Slowdown and Acceleration." *Science* 345 (2014): 897–903.

Drijfhout, S. S., A. T. Blaker, S. A. Josey, A.J.G. Nurser, B. Sinha, and M. A. Balmaseda. "Surface Warming Hiatus Caused by Increased Heat Uptake across Multiple Ocean Basins." *Geophysical Research Letters* 41 (2014): 7868–74.

Durack, P. J., P. J. Gleckler, F. W. Landerer, and K. E. Taylor. "Quantifying Underestimates of Long-Term Upper-Ocean Warming." *Nature Climate Change* 4 (2014): 999–1005.

GISTEMP Team. "GISS Surface Temperature Analysis (GISTEMP)." *NASA Goddard Institute for Space Studies*. Accessed February 7, 2017. http://data.giss.nasa.gov/gistemp/.

"Global Surface Temperature." *Met Office*. Accessed February 7, 2017. http://www.metoffice.gov.uk/research/monitoring/climate/surface -temperature.

Hobbs, W. R., and J. K. Willis. "Detection of an Observed 135-Year Ocean Temperature Change from Limited Data." *Geophysical Research Letters* 40 (2013): 2252–58.

Llovel, W., J. K. Willis, F. W. Landerer, and I. Fukumori. "Deep-Ocean Contribution to Sea Level and Energy Budget Not Detectable over the Past Decade." *Nature Climate Change* 4 (2014): 1031–35.

Lyman J. M., S. A. Good, V. V. Gouretski, M. Ishii, G. C. Johnson, M. D. Palmer, D. M. Smith, and J. K. Willis. "Robust Warming of the Global Upper Ocean." *Nature* 465 (2010): 334–37.

Roemmich, D., W. J. Gould, and J. Gilson. "135 Years of Global Ocean Warming between the Challenger Expedition and the Argo Programme." *Nature Climate Change* 2 (2012): 425–28.

"Summary of Findings." *Berkeley Earth*. Accessed February 7, 2017. http://berkeleyearth.org/summary-of-findings/.

KEY SOURCES FOR CHAPTER 2

Abramov, O., and S. J. Mojzsis. "Microbial Habitability of the Hadean Earth during the Late Heavy Bombardment." *Nature* 459 (2009): 419–22.

Albarède, F. "Volatile Accretion History of the Terrestrial Planets and Dynamic Implications." *Nature* 461 (2009): 1227–33.

Altwegg, K., H. Balsiger, A. Bar-Nun, J. J. Berthelier, A. Bieler, P. Bochsler, C. Briois, et al. "67P/Churyumov-Gerasimenko, a Jupiter Family Comet with a High D/H Ratio." *Science* 347 (2015): 1261952. doi:10.1126/science.1261952.

Amante, C., and B. W. Eakins (2009). *ETOPO1 1 Arc-Minute Global Relief Model: Procedures, Data Sources and Analysis*. NOAA Technical Memorandum NESDIS NGDC-24. N.p.: National Geophysical

Data Center, National Oceanic and Atmospheric Administration, 2009. doi:10.7289/V5C8276M.

Bhattacharya, D., and L. Medlin. "Algal Phylogeny and the Origin of Land Plants." *Plant Physiology* 116 (1998): 9–15.

Bieler, A., K. Altwegg, H. Balsinger, A. Bar-Nun, J.-J. Berthelier, P. Bochsler, C. Briois, et al. (2015). "Abundant Molecular Oxygen in the Coma of Comet 67P/Churyumov-Gerasimenko." *Nature* 526 (2015): 678–81.

Biello, D. "The Origin of Oxygen in Earth's Atmosphere." *Scientific American*, August 19, 2009, http://www.scientificamerican.com /article/origin-of-oxygen-in-atmosphere/.

Cleeves, L. I., E. A. Bergin, C.M.O'D. Alexander, F. Du, D. Graninger, K. I. Öberg, and T. J. Harries. "The Ancient Heritage of Water Ice in the Solar System." *Science* 345 (2014): 1590–93.

Connelly, J. N., M. Bizzarro, A. N. Krot, A. Nordlund, D. Wielandt, and M. A. Ivanova. "The Absolute Chronology and Thermal Processing of Solids in the Solar Protoplanetary Disk." *Science* 338 (2012): 651–55.

Cooper, G. M. *The Cell: A Molecular Approach.* 2nd ed. Sunderland, MA: Sinauer Associates, 2000. http://www.ncbi.nlm.nih.gov/books /NBK9841/.

Daeschler, E. B., N. H. Shubin, and F. A. Jenkins Jr. "A Devonian Tetrapod-like Fish and the Evolution of the Tetrapod Body Plan." *Nature* 440 (2006): 757–63.

Elkins-Tanton, L. T. "Formation of Early Water Oceans on Rocky Planets." *Astrophysics and Space Science* 332 (2010): 359–64.

Endal, A. S., and S. Sofia. "The Evolution of Rotating Stars, I: Method and Exploratory Calculations for a 7 MQ Star." *Astrophysical Journal* 210 (1981): 184–98.

Goesmann, F., H. Rosenbauer, J. H. Bredehoft, M. Cabane, P. Ehrenfreund, T. Gautier, C. Giri, et al. "Organic Compounds on Comet 67P/Churyumov-Gerasimenko Revealed by COSAC Mass Spectrometry." *Science* 349 (2015): aab0689. doi:10.1126/science.aab0689.

Gvirtzman, Z., and R. J. Stern. "Bathymetry of Mariana Trench-Arc System and Formation of the Challenger Deep as a Consequence of Weak Plate Coupling." *Tectonics* 23 (2004): TC2011. doi:10.1029/2003TC001581.

Han, T. M., and B. Runnegar. "Megascopic Eukaryotic Algae from the 2.1-Billion-Year-Old Negaunee Iron-Formation, Michigan." *Science* 257 (1992): 232–35.

Hartogh, P., D. C. Lis, D. Bockelée-Morvan, M. de Val-Borro, N. Biver, M. Küppers, M. Emprechtinger, et al. "Ocean-like Water in the Jupiter-Family Comet 103P/Hartley 2." *Nature* 478 (2011): 218–20.

Holland, G., C. J. Ballentine, and M. Cassidy. "Meteorite Kr in Earth's Mantle Suggests a Late Accretionary Source for the Atmosphere." *Science* 326 (2009): 1522–25.

Hsu, K., H. Oberhänsli, J. Y. Gao, S. Sun, H. Chen, and U. Krähenbühl. " 'Strangelove Ocean' before the Cambrian Explosion." *Nature* 316 (1985): 809–11.

Kasting, J. F., and M. T. Howard. "Atmospheric Composition and Climate on the Early Earth." *Philosophical Transactions of the Royal Society of London, B* 361 (2006): 1733–42.

Konhauser, K. O., S. V. Lalonde, N. J. Planavsky, E. Pecoits, T. W. Lyons, S. J. Mojzsis, O. J. Rouxel, et al. "Aerobic Bacterial Pyrite Oxidation and Acid Rock Drainage during the Great Oxidation Event." *Nature* 478 (2011): 369–73.

Leslie, M. "On the Origin of Photosynthesis." *Science* 323 (2009): 1286–87.

Lyons, T. W., C. T. Reinhard, and N. J. Planavsky. "The Rise of Oxygen in Earth's Early Ocean and Atmosphere." *Nature* 506 (2014): 307–15.

MacNaughton, R. B., J. M. Cole, R. W. Dalrymple, S. J. Braddy, D.E.G. Briggs, and T. D. Lukie. "First Steps on Land: Arthropod Trackways in Cambrian-Ordovician Eolian Sandstone, Southeastern Ontario, Canada." *Geology* 30 (2002): 391–94.

Marchi, S., C. R. Chapman, C. I. Fassett, J. W. Head, W. F. Bottke, and R. G. Strom. "Global Resurfacing of Mercury 4.0–4.1 Billion Years Ago by Heavy Bombardment and Volcanism." *Nature* 499 (2013): 59–61.

Müller, R. D., M. Sdrolias, C. Gaina, and W. R. Roest. "Age, Spreading Rates and Spreading Symmetry of the World's Ocean Crust." *Geochemistry Geophysics Geosystems* 9 (2008): Q04006. doi:10.1029 /2007GC001743.

Nutman, A. P., V. C. Bennett, C.R.L. Friend, M. J. van Kranendonk, and A. R. Chivas. "Rapid Emergence of Life Shown by Discovery of 3,700-Million-Year-Old Microbial Structures." *Nature* 537 (2016): 535–38. doi:10.1038/nature19355.

Olsen, J. M. "Photosynthesis in the Archean Era." *Photosynthesis Research* 88 (2006): 109–17.

Oren, A., E. Padan, and M. Avron. "Quantum Yields for Oxygenic and Anoxygenic Photosynthesis in the Cyanobacterium *Oscillatoria limnetica*." *Proceedings of the National Academy of Sciences of the USA* 74 (1977): 2152–56.

"Oxygen Solubility in Fresh Water and Sea Water." *The Engineering ToolBox*. Accessed February 8, 2017. http://www.engineeringtool box.com/oxygen-solubility-water-d_841.html.

Patel, B. H., C. Percivalle, D. J. Ritson, C. D. Duffy, and J. D. Sutherland. "Common Origins of RNA, Protein and Lipid Precursors in a Cyanosulfidic Protometabolism." *Nature Chemistry* 7 (2015): 301–7.

Quanz, S. P., A. Amara, M. R. Meyer, J. H. Girard, M. A. Kenworthy, and M. Kasper. "Confirmation and Characterization of the Protoplanet HD 100546 b: Direct Evidence for Gas Giant Planet Formation at 50 AU." *Astrophysical Journal* 807 (2015): 64. doi:10.1088 /0004-637X/807/1/64.

Rahmstorf, S. "Thermohaline Ocean Circulation." In *Encyclopedia of Quaternary Sciences*, edited by S. A. Elias. Amsterdam: Elsevier, 2006.

Roemmich, D., W. J. Gould, and J. Gilson. "135 Years of Global Ocean Warming between the Challenger Expedition and the Argo Programme." *Nature Climate Change* 2 (2012): 425–28.

Sarafian, A. R., S. G. Nielsen, H. R. Marschall, F. M. McCubbin, and B. D. Monteleone. "Early Accretion of Water in the Inner Solar System from a Carbonaceous Chondrite-like Source." *Science* 346 (2014): 623–26.

Satkoski, A. M., N. J. Beukes, W. Li, B. L. Beard, and C. M. Johnson. "A Redox-Stratified Ocean 3.2 Billion Years Ago." *Earth and Planetary Science Letters* 430 (2015): 43–53.

Scotese, C. R. "PALEOMAP Project." *Scotese.com*. Accessed February 8, 2017. http://www.scotese.com/earth.htm.

Shen, B., L. Dong, S. Xiao, and M. Kowalewski. "The Avalon Explosion: Evolution of Ediacara Morphospace." *Science* 319 (2008): 81–84.

Sigman, D. M., and M. P. Hain. "The Biological Productivity of the Ocean." *Nature Education Knowledge* 3 (2012): 1–16.

Tarduno, J. A., R. D. Cottrell, W. J. Davis, F. Nimmo, and R. K. Bono. "A Hadean to Paleoarchean Geodynamo Recorded by Single Zircon Crystals." *Science* 349 (2015): 521–24.

USGS. "Mid-ocean Ridges." *Encyclopedia of Earth*. Last modified October 18, 2016. http://www.eoearth.org/view/article/164696/.

Valley, J. W., A. J. Cavosie, T. Ushikubo, D. A. Reinhard, D. F. Lawrence, D. J. Larson, P. H. Clifton, et al. "Hadean Age for a Post-Magma-Ocean Zircon Confirmed by Atom-Probe Tomography." *Nature Geoscience* 7 (2014): 219–23.

"When Did Eukaryotic Cells (Cells with Nuclei and Other Internal Organelles) First Evolve? What Do We Know about How They Evolved from Earlier Life-Forms?" *Scientific American*. Accessed February 8, 2017. http://www.scientificamerican.com/article/when-did-eukaryotic-cells/.

Worthington, L. V. (1981). "The Water Masses of the World Ocean: Some Results of a Finite-Scale Census." In *Evolution of Physical Oceanography*, edited by B. A. Warren and C. Wunsch, 42–69. Cambridge, MA: MIT Press.

Zhu, S., M. Zhu, A. H. Knoll, Z. Yin, F. Zhao, S. Sun, Y. Qu, M. Shi, and H. Liu. "Decimetre-Scale Multicellular Eukaryotes from the 1.56-Billion-Year-Old Gaoyuzhuang Formation in North China." *Nature Communications* 7 (2016): 11500. doi:10.1038/ncomms11500.

KEY SOURCES FOR CHAPTER 3

Anbar, A. D., Y. Duan, T. W. Lyons, G. L. Arnold, B. Kendall, R. A. Creaser, A. J. Kaufman, et al. "A Whiff of Oxygen before the Great Oxidation Event?" *Science* 317 (2007): 1903–6.

Berner, R. A. "Atmospheric Oxygen over Phanerozoic Time." *Proceedings of the National Academy of Sciences of the USA* 96 (1999): 10955–57.

"Climate Change Impacts on Methane Hydrates." *World Ocean Review*. Accessed February 8, 2017. http://worldoceanreview.com/en/comments/feed/wor-1/ocean-chemistry/climate-change-and-methane-hydrates/.

Dutkiewicz, A., H. Volk, S. George, J. Ridley, and R. Buick. "Biomarkers from Huronian Oil-Bearing Fluid Inclusions: An Uncontaminated Record of Life before the Great Oxidation Event." *Geology* 34 (2006): 437–40.

Garric, G., and M. Huber. "Quasi-Decadal Variability in Paleoclimate Records: Sunspot Cycles or Intrinsic Oscillations?" *Paleoceanography* 18 (2003): PA1068. doi:10.1029/2002PA000869.

"Gas Hydrate: What Is It?" *USGS Woods Hole Science Center* at *web.archive.org*. Last modified August 31, 2009. http://web.archive.org/web/20120614141539/http://woodshole.er.usgs.gov/project-pages/hydrates/what.html.

Hester, K. C., and P. G. Brewer. "Clathrate Hydrates in Nature." *Annual Reviews in Marine Science* 1 (2009): 303–27.

Hoffmann, R. "Old Gas, New Gas: Methane—Made and Taken Apart by Microbes, in the Earth, by People." *American Scientist* 94 (2006): 16. http://www.americanscientist.org/issues/pub/2006/1/old-gas-new-gas.

Holland, H. D. "The Oxygenation of the Atmosphere and Oceans." *Philosophical Transactions of the Royal Society of London, B* 361 (2006): 903–15.

KamLAND Collaboration. "Partial Radiogenic Heat Model for Earth Revealed by Geoneutrino Measurements." *Nature Geoscience* 4 (2011): 647–51.

"Large Igneous Province." *Wikipedia.* Accessed February 8, 2017. https://en.wikipedia.org/wiki/Large_igneous_province.

Milkov, A. V. "Global Estimates of Hydrate-Bound Gas in Marine Sediments: How Much Is Really Out There? *Earth Science Reviews* 66 (2004): 183–97.

Omori, S., and M. Santosh. "Metamorphic Decarbonation in the Neoproterozoic and Its Environmental Implication." *Gondwana Research* 14 (2008): 97–104.

Peixoto, J. P., and M. A. Kettani. "The Control of the Water Cycle." *Scientific American* 228, no. 4 (1973): 46–61.

Pollack, H. N., S. J. Hurter, and J. R. Johnson. "Heat Flow from the Earth's Interior: Analysis of the Global Data Set." *Reviews of Geophysics* 31 (1993): 267–80.

Rosing, M. T., D. K. Bird, N. H. Sleep, and C. J. Bjerrum. "No Climate Paradox under the Faint Early Sun." *Nature* 464 (2010): 744–47.

Royer, D. L. "CO_2-Forced Climate Thresholds during the Phanerozoic." *Geochimica et Cosmochimica Acta* 70 (2006): 5665–75.

Santosh, M., and S. Omori. "CO_2 Flushing: A Plate Tectonic Perspective." *Gondwana Research* 13 (2008): 86–102.

"Snowball Events in Earth History." *Snowball Earth.* Accessed February 8, 2017. http://www.snowballearth.org/events.html.

Sonett, C. P., E. P. Kvale, A. Zakharian, M. A. Chan, and T. M. Demko. "Late Proterozoic and Paleozoic Tides, Retreat of the Moon, and Rotation of the Earth." *Science* 273 (1996): 100–104.

Steemans, P., L. Herisse, J. Melvin, A. Miller, F. Paris, J. Verniers, and H. Wellman. "Origin and Radiation of the Earliest Vascular Land Plants." *Science* 324 (2009): 353.

Tang, H., and Y. Chen. "Global Glaciations and Atmospheric Change at ca. 2.3 Ga." *Geoscience Frontiers* 4 (2013): 583–96.

Vellekoop, J., A. Sluijs, J. Smit, S. Schouten, J.W.H. Weijers, J. S. Sinninghe Damsté, and H. Brinkhuis. "Rapid Short-Term Cooling following the Chicxulub Impact at the Cretaceous-Paleogene Boundary." *Proceedings of the National Academy of Sciences of the USA* 111 (2014): 7537–41.

Williams, G. E. "Precambrian Length of Day and the Validity of Tidal Rhythmite Paleotidal Values." *Geophysical Research Letters* 24 (1997): 421–24.

Young, G. M. "Precambrian Supercontinents, Glaciations, Atmospheric Oxygenation, Metazoan Evolution and an Impact That May Have Changed the Second Half of Earth History." *Geoscience Frontiers* 4 (2013): 247–61.

KEY SOURCES FOR CHAPTER 4

Broecker, W. S., and T. -H. Peng. "The Role of $CaCO_3$ Compensation in the Glacial to Interglacial Atmospheric CO Change." *Global Biogeochemical Cycles* 1 (1987): 15–29.

Budd, G. E. "The Cambrian Fossil Record and the Origin of the Phyla." *Integrative and Comparative Biology* 43 (2003): 157–65.

Budd, G. E. "At the Origin of Animals: The Revolutionary Cambrian Fossil Record." *Current Genomics* 14 (2013): 344–54.

Budd, G. E., and S. Jensen. "A Critical Reappraisal of the Fossil Record of the Bilaterian Phyla." *Biological Reviews* 75 (2007): 253–95.

Butterfield, N. J. "Hooking Some Stem-Group 'Worms': Fossil Lopho-trochozoans in the Burgess Shale." *Bioessays* 28 (2006): 1161–66.

"The Cambrian Explosion." *Burgess Shale.* Accessed February 8, 2017. http://burgess-shale.rom.on.ca/en/science/origin/04-cambrian -explosion.php.

Conway-Morris, S. "The Cambrian 'Explosion' of Metazoans and Molecular Biology: Would Darwin be Satisfied?" *International Journal of Developmental Biology* 47 (2003): 505–15.

Crowley, T. J., W. T. Hyde, and W. R. Peltier. "CO_2 Levels Required for Deglaciation of a 'Near-Snowball' Earth." *Geophysical Research Letters* 28 (2001): 283–86.

Dutch, S. "Plate Tectonics and Earth History." *University of Wisconsin—Green Bay.* Last modified December 4, 2009. https://www .uwgb.edu/dutchs/EarthSC102Notes/102PTEarthHist.htm.

Erwin, D. H., and E. H. Davidson. "The Last Common Bilaterian Ancestor." *Development* 129 (2002): 3021–32.

Erwin, D. H., M. Laflamme, S. M. Tweedt, E. A. Sperling, D. Pisani, and K. J. Peterson. "The Cambrian Conundrum: Early Divergence and Later Ecological Success in the Early History of Animals." *Science* 334 (2011): 1091–97.

Eyles, N., and N. Januszczak. "'Zipper-Rift': A Tectonic Model for Neoproterozoic Glaciations during the Breakup of Rodinia after 750 Ma." *Earth-Science Reviews* 65 (2004): 1–73.

Ferrari, R. "What Goes Down Must Come Up." *Nature* 513 (2014): 179–80.

Feulner, G., C. Hallmann, and H. Kienert. "Snowball Cooling after Algal Rise." *Nature Geoscience* 8 (2015): 659–62.

Ganade de Araujo, C. E., D. Rubatto, J. Hermann, U. G. Cordani, R. Caby, and M. A. Basei. "Ediacaran 2,500-km-long Synchronous Deep Continental Subduction in the West Gondwana Orogen." *Nature Communications* 5 (2014): 5198. doi:10.1038/ncomms6198.

Gernon, T. M., T. K. Hincks, T. Tyrrell, E. J. Rohling, and M. R. Palmer. "Snowball Earth Ocean Chemistry Driven by Extensive Ridge Volcanism during Rodinia Breakup." *Nature Geoscience* 9 (2016): 242–48.

Hoffman, P. F., A. J. Kaufman, G. P. Halverson, and D. P. Schrag. "A Neoproterozoic Snowball Earth." *Science* 281 (1998): 1342–46.

Jacobsen, S. B. "Earth Science: Gas Hydrates and Deglaciations." *Nature* 412 (2001): 691–93.

Kerr, R. A. "Early Life Thrived despite Earthly Travails." *Science* 284 (1999): 2111–13.

Knoll, A. H., and S. B. Carroll. "Early Animal Evolution: Emerging Views from Comparative Biology and Geology." *Science* 284 (1999): 2129–37.

Kouchinsky, A., S. Bengtson, B. Runnegar, C. Skovsted, M. Steiner, and M. Vendrasco. "Chronology of Early Cambrian Biomineralization." *Geological Magazine* 149 (2012): 221–51.

Lenton, T. M., R. A. Boyle, S. W. Poulton, G. A. Shields-Zhou, and N. J. Butterfield. "Co-evolution of Eukaryotes and Ocean Oxygenation in the Neoproterozoic Era." *Nature Geoscience* 7 (2014): 257–65.

Li, C. W., J. Y. Chen, and T. E. Hua. "Precambrian Sponges with Cellular Structures." *Science* 279 (1998): 879–82.

Li, Z. X., S. V. Bogdanova, A. S. Collins, A. Davidson, B. De Waele, R. E. Ernst, I.C.W. Fitzsimons, et al. "Assembly, Configuration, and Break-Up History of Rodinia: A Synthesis." *Precambrian Research* 160 (2008): 179–210.

Lieberman, B. S. "Testing the Darwinian Legacy of the Cambrian Radiation Using Trilobite Phylogeny and Biogeography." *Journal of Paleontology* 73 (1999): 176–81.

Maas, A., A. Braun, X.-P. Dong, P.C.J. Donoghue, K. J. Müller, E. Olempska, J. E. Repetski, D. J. Sivester, M. Stein, and D. Waloszek (2006). "The 'Orsten': More than a Cambrian Konservat-Lagerstätte Yielding Exceptional Preservation." *Palaeoworld* 15 (2006): 266–82.

Macdonald, F. A., M. D. Schmitz, J. L. Crowley, C. F. Roots, D. S. Jones, A. C. Maloof, J. V. Strauss, P. A. Cohen, D. T. Johnston, and D. P. Schrag. "Calibrating the Cryogenian." *Science* 327 (2010): 1241–43.

Maloof, A. C., S. M. Porter, D. A. Fike, and M. P. Eddy. "The Earliest Cambrian Record of Animals and Ocean Geochemical Change." *Geological Society of America Bulletin* 122 (2010): 1731–74.

McMenamin, M. A. "Ediacaran Biota from Sonora, Mexico." *Proceedings of the National Academy of Sciences of the USA* 93 (1996): 4990–93.

Meert, J. G., A. S. Gibsher, N. M. Levashova, W. C. Grice, G. D. Kamenov, and A. B. Ryabinin. "Glaciation and ~770 Ma Ediacara

(?) Fossils from the Lesser Karatau Microcontinent, Kazakhstan." *Gondwana Research* 19 (2011): 867–80.

Melezhik, V. "Multiple Causes of Earth's Earliest Global Glaciation." *Terra Nova* 18 (2006): 130–37.

Misra, S. "Late Precambrian (?) Fossils from Southeastern Newfoundland." *Geological Society of America Bulletin* 80 (1969): 2133–40.

Morse, J. W., R. S. Arvidson, and A. Lüttge. "Calcium Carbonate Formation and Dissolution." *Chemical Reviews* 107 (2007): 342–81.

Narbonne, G. M. "The Ediacara Biota: A Terminal Neoproterozoic Experiment in the Evolution of Life." *Geological Society of America Today* 8 (1998): 1–6.

Och, L., and G. Shields-Zhou. "The Neoproterozoic Oxygenation Event: Environmental Perturbations and Biogeochemical Cycling." *Earth-Science Reviews* 110 (2012): 26–57.

Pierrehumbert, R. T. "High Levels of Atmospheric Carbon Dioxide Necessary for the Termination of Global Glaciation." *Nature* 429 (2004): 646–49.

Planavsky, N. J., O. J. Rouxel, A. Bekker, S. V. Lalonde, K. O. Konhauser, C. T. Reinhard, and T. W. Lyons. "The Evolution of the Marine Phosphate Reservoir." *Nature* 467 (2010): 1088–90.

Rayner-Canham, G., and J. Grandy. "Did Molybdenum Control Evolution on Earth?" *Education in Chemistry* September 2011: 144–47.

Ridgwell, A., and R. E. Zeebe. "The Role of the Global Carbonate Cycle in the Regulation and Evolution of the Earth System." *Earth and Planetary Science Letters* 234 (2005): 299–315.

Rothman, D. H., J. M. Hayes, and R. E. Summons. "Dynamics of the Neoproterozoic Carbon Cycle." *Proceedings of the National Academy of Sciences of the USA* 100 (2003): 8124–29.

Schrag, D., R. Berner, P. Hoffman, and G. Halverson. "On the Initiation of a Snowball Earth." *Geochemistry Geophysics Geosystems* 3 (2002): 1–21.

Sheldon, N. "Precambrian Paleosols and Atmospheric CO_2 Levels." *Precambrian Research* 147 (2006): 148–55.

Shen, B., L. Dong, S. Xiao, and M. Kowalewski. "The Avalon Explosion: Evolution of Ediacara Morphospace." *Science* 319 (2008): 81–84.

Shields-Zhou, G., and L. Och. "The Case for a Neoproterozoic Oxygenation Event: Geochemical Evidence and Biological Consequences." *Geological Society of America Today* 21 (2011): 4–11.

Snowball Earth. Accessed February 8, 2017. http://www.snowballearth .org/.

Tang, H., and Y. Chen. "Global Glaciations and Atmospheric Change at ca. 2.3 Ga." *Geoscience Frontiers* 4 (2013): 583–96.

Valentine, J. W., D. Jablonski, and D. H. Erwin. "Fossils, Molecules and Embryos: New Perspectives on the Cambrian Explosion." *Development* 126 (1999): 851–59.

Yin, L., S. Xiao, and X. Yuan. "New Observations on Spiculelike Structures from Doushantuo Phosphorites at Weng'an, Guizhou Province." *Chinese Science Bulletin* 46 (2001): 1828–32.

Yin, L., M. Zhu, A. H. Knoll, X. Yuan, J. Zhang, and J. Hu. "Doushantuo Embryos Preserved Inside Diapause Egg Cysts." *Nature* 446 (2007): 661–63.

Young, G. M. "Precambrian Supercontinents, Glaciations, Atmospheric Oxygenation, Metazoan Evolution and an Impact That May Have Changed the Second Half of Earth History." *Geoscience Frontiers* 4 (2013): 247–61.

KEY SOURCES FOR CHAPTER 5

Alexander, K., K. J. Meissner, and T. J. Bralower. "Sudden Spreading of Corrosive Bottom Water during the Palaeocene-Eocene Thermal Maximum." *Nature Geoscience* 8 (2015): 458–61.

Berger, W., and P. Roth. "Oceanic Micropaleontology: Progress and Prospect." *Reviews of Geophysics* 13 (1975): 561–85.

Brennecka, G. A., A. D. Herrmann, T. J. Algeo, and A. D. Anbar. "Rapid Expansion of Oceanic Anoxia Immediately Before the End-Permian Mass Extinction." *Proceedings of the National Academy of Sciences of the USA* 108 (2011): 17631–34.

Briggs, D., E. Clarkson, and R. Aldridge. "The Conodont Animal." *Lethaia* 16 (2007): 1–14.

Burgess, S. D., S. Bowring, and S. Z. Shen. "High-Precision Timeline for Earth's Most Severe Extinction." *Proceedings of the National Academy of Sciences of the USA* 111 (2014): 3316–21.

Clarkson, M. O., S. A. Kasemann, R. A. Wood, T. M. Lenton, S. J. Daines, S. Richoz, F. Ohnemueller, A. Meixner, S. W. Poulton, and E. T. Tipper. "Ocean Acidification and the Permo-Triassic Mass Extinction." *Science* 348 (2015): 229–32.

"Climate Change 2007: Working Group I: The Physical Science Basis. Frequently Asked Question 6.1: What Caused the Ice Ages and Other Important Climate Changes before the Industrial Era?" *Intergovernmental Panel on Climate Change.* Accessed February 8, 2017. https://www.ipcc.ch/publications_and_data/ar4/wg1/en/faq -6-1.html.

Cui, Y., L. Kump, A. Ridgwell, A. Charles, C. Junium, A. Diefendorf, K. Freeman, N. Urban, and I. Harding. "Slow Release of Fossil Carbon during the Palaeocene–Eocene Thermal Maximum." *Nature Geoscience* 4 (2011): 481–85.

DeConto, R. M., S. Galeotti, M. Pagani, D. Tracy, K. Schaefer, T. Zhang, D. Pollard, and D. J. Beerling. "Past Extreme Warming Events Linked to Massive Carbon Release from Thawing Permafrost." *Nature* 484 (2012): 87–91.

Dickens, G. R., M. M. Castillo, and J. C. Walker. "A Blast of Gas in the Latest Paleocene: Simulating First-Order Effects of Massive Dissociation of Oceanic Methane Hydrate." *Geology* 25 (1997): 259–62.

Dickens, G. R., J. O'Neil, D. Rea, and R. Owen. "Dissociation of Oceanic Methane Hydrate as a Cause of the Carbon Isotope Excursion at the End of the Paleocene." *Paleoceanography* 10 (1995): 965–71.

Feely, R. A., C. L. Sabine, K. Lee, W. Berelson, J. Kleypas, V. J. Fabry, and F. J. Millero. "Impact of Anthropogenic CO_2 on the $CaCO_3$ System in the Oceans." *Science* 305 (2004): 362–66.

Grice, K., C. Cao, G. D. Love, M. E. Böttcher, R. J. Twitchett, E. Grosjean, R. E. Summons, S. C. Turgeon, W. Dunning, and Y. C. Jin. "Photic Zone Euxinia during the Permian-Triassic Superanoxic Event." *Science* 307 (2005): 706–9.

Hain, M. P., D. M. Sigman, and G. H. Haug. "The Biological Pump in the Past." In *Treatise on Geochemistry*, vol. 8, edited by M. J. Mottl and H. Elderfield, 2nd ed., 485–517. Amsterdam: Elsevier, 2014.

Hönisch, B., A. Ridgwell, D. N. Schmidt, E. Thomas, S. J. Gibbs, A. Sluijs, R. Zeebe, et al. "The Geological Record of Ocean Acidification." *Science* 335 (2012): 1058–63.

Jardine, P. "Patterns in Palaeontology: The Paleocene-Eocene Thermal Maximum." *Palaeontology[online]* 1 (2011): 5. http://www.palaeon tologyonline.com/articles/2011/the-paleocene-eocene-thermal -maximum/.

Lehrmann, D., J. Ramezani, S. Bowring, M. Martin, P. Montgomery, P. Enos, J. Payne, M. Orchard, W. Hongmei, and W. Jiayong. "Timing of Recovery from the End-Permian Extinction: Geochronologic and Biostratigraphic Constraints from South China." *Geology* 34 (2006): 1053–56.

Meissner, K., T. Bralower, K. Alexander, T. Jones, W. Sijp, and M. Ward. "The Paleocene-Eocene Thermal Maximum: How Much Carbon Is Enough?" *Paleoceanography* 29 (2014): 946–63.

Notman, N. "The Other Carbon Dioxide Problem." *Chemistry World* July 29, 2014, http://www.rsc.org/chemistryworld/2014/07/ocean -acidification.

Ocean Portal Team. "Ocean Acidification." *Ocean Portal.* Accessed February 8, 2017. http://ocean.si.edu/ocean-acidification.

Pälike, H., M. W. Lyle, H. Nishi, I. Raffi, A. Ridgwell, K. Gamage, A. Klaus, et al. "A Cenozoic Record of the Equatorial Pacific Carbonate Compensation Depth." *Nature* 488 (2012): 609–14.

Raven, J., and P. Falkowski. "Oceanic Sinks for Atmospheric CO_2." *Plant, Cell & Environment* 22 (1999): 741–55.

Ridgwell, A., and R. E. Zeebe. "The Role of the Global Carbonate Cycle in the Regulation and Evolution of the Earth System." *Earth and Planetary Science Letters* 234 (2005): 299–315.

Rohde, R. A. "File: Phanerozoic Carbon Dioxide.png." *Wikipedia.* Last modified February 25, 2006. https://en.wikipedia.org/wiki /File:Phanerozoic_Carbon_Dioxide.png.

Sahney, S., and M. J. Benton. "Recovery from the Most Profound Mass Extinction of All Time." *Proceedings of the Royal Society of London, B* 275 (2008): 759–65.

Schobben, M., A. Stebbins, A. Ghaderi, H. Strauss, D. Korn, and C. Korte. "Flourishing Ocean Drives the End-Permian Marine Mass

Extinction." *Proceedings of the National Academy of Sciences of the USA* 112 (2015): 10298–303.

Schubert, B. A., and A. Hope Jahren. "Reconciliation of Marine and Terrestrial Carbon Isotope Excursions Based on Changing Atmospheric CO_2 Levels." *Nature Communications* 4 (2013): 1653. doi:10.1038/ncomms2659.

Shu, D.-G., H.-L. Luo, C. S. Morris, X.-L. Zhang, S.-X. Hu, L. Chen, J. Han, M. Zhu, Y. Li, and L.-Z. Chen. "Lower Cambrian Vertebrates from South China." *Nature* 402 (1999): 42–46.

Sluijs, A. "Carbon Burp and Transient Global Warming during the Paleocene-Eocene Thermal Maximum." *PAGES News* 16 (2008): 9–11.

Sluijs, A., S. Schouten, M. Pagani, M. Woltering, H. Brinkhuis, J. S. Sinninghe Damsté, G. R. Dickens, et al. "Subtropical Arctic Ocean Temperatures during the Palaeocene/Eocene Thermal Maximum." *Nature* 441 (2006): 610–13.

Sun, Y., M. M. Joachimski, P. B. Wignall, C. Yan, Y. Chen, H. Jiang, L. Wang, and X. Lai. "Lethally Hot Temperatures during the Early Triassic Greenhouse." *Science* 338 (2012): 366–70.

Svensen, H., S. Planke, A. Malthe-Sørenssen, B. Jamtveit, R. Myklebust, T. Rasmussen Eidem, and S. S. Rey. "Release of Methane from a Volcanic Basin as a Mechanism for Initial Eocene Global Warming." *Nature* 429 (2004): 542–45.

Van Andel, T. H., G. R. Heath, and T. C. Moore Jr. "Cenozoic History and Paleoceanography of the Central Equatorial Pacific Ocean: A Regional Synthesis of Deep Sea Drilling Project Data." *Geological Society of America* 143 (1975): 1–134.

Van Andel, T. H., J. Thiede, J. G. Sclater, and W. W. Hay. "Depositional History of the South Atlantic Ocean during the Last 125 Million Years." *Journal of Geology* 85 (1977): 651–98.

"Vertebrates: Fossil Record." *University of California Museum of Paleontology*. Last modified February 2, 2005. http://www.ucmp.berkeley.edu/vertebrates/vertfr.html.

"What Is Ocean Acidification?" *PMEL Carbon Program*. Accessed February 8, 2017. http://www.pmel.noaa.gov/co2/story/What+is+Ocean+Acidification%3F.

Zachos, J. C., G. R. Dickens, and R. E. Zeebe. "An Early Cenozoic Perspective on Greenhouse Warming and Carbon-Cycle Dynamics." *Nature* 451 (2008): 279–83.

Zachos, J. C., M. Pagani, L. Sloan, E. Thomas, and K. Billups. "Trends, Rhythms, and Aberrations in Global Climate 65 Ma to Present." *Science* 292 (2001): 686–93.

Zachos, J. C., U. Röhl, S. A. Schellenberg, A. Sluijs, D. A. Hodell, D. C. Kelly, E. Thomas, et al. "Rapid Acidification of the Ocean during the Paleocene-Eocene Thermal Maximum." *Science* 308 (2005): 1611–15.

Zeebe, R. E. "History of Seawater Carbonate Chemistry, Atmospheric CO_2, and Ocean Acidification." *Annual Review of Earth and Planetary Science* 40 (2012): 141–65.

Zeebe, R. E., A. Ridgwell, and J. C. Zachos. "Anthropogenic Carbon Release Rate Unprecedented during the Past 66 Million Years." *Nature Geoscience* 9 (2016): 325–29.

Zeebe, R. E., J. C. Zachos, and G. R. Dickens. "Carbon Dioxide Forcing Alone Insufficient to Explain Palaeocene–Eocene Thermal Maximum Warming." *Nature Geoscience* 2 (2009): 576–80.

KEY SOURCES FOR CHAPTER 6

Archer, D., P. Martin, B. Buffett, V. Brovkin, S. Rahmstorf, and A. Ganopolski. "The Importance of Ocean Temperature to Global Biogeochemistry." *Earth and Planetary Science Letters* 222 (2004): 333–48.

Bache, F., J. Gargani, J.-P. Suc, C. Gorini, M. Rabineau, S. Popescu, E. Leroux, et al. "Messinian Evaporite Deposition during Sea Level Rise in the Gulf of Lions (Western Mediterranean)." *Marine and Petroleum Geology* 66 (2015): 262–77.

Blackburn, T. J., P. E. Olsen, S. A. Bowring, N. M. McLean, D. V. Kent, J. Puffer, G. McHone, E. T. Rasbury, and M. Et-Touhami. "Zircon U-Pb Geochronology Links the End-Triassic Extinction with the Central Atlantic Magmatic Province." *Science* 340 (2013): 941–45.

Breecker, D. O., Z. D. Sharp, and L. D. McFadden. "Atmospheric CO_2 Concentrations during Ancient Greenhouse Climates Were Similar to Those Predicted for A.D. 2100." *Proceedings of the National Academy of Sciences of the USA* 107 (2010): 576–80.

Chaboureau, A., Y. Donnadieu, P. Sepulchre, C. Robin, F. Guillocheau, and S. Rohais. "The Aptian Evaporites of the South Atlantic: A Climatic Paradox?" *Climate of the Past* 8 (2012): 1047–58.

CIESM. *The Messinian Salinity Crisis from Mega-Deposits to Microbiology: A Consensus Report.* CIESM Workshop Monographs 33. Edited by F. Briand. Monaco: CIESM, 2008.

"Dissolved Oxygen." *Fondriest Environmental, Inc.* Last modified November 19, 2013. http://www.fondriest.com/environmental -measurements/parameters/water-quality/dissolved-oxygen/.

Elderfield, H., P. Ferretti, M. Greaves, S. Crowhurst, I. N. McCave, D. Hodell, and A. M. Piotrowski. "Evolution of Ocean Temperature and Ice Volume through the Mid-Pleistocene Climate Transition." *Science* 337 (2012): 704–9.

Emeis, K.-C., and H. Weissert. "Tethyan-Mediterranean Organic Carbon-Rich Sediments from Mesozoic Black Shales to Sapropels." *Sedimentology* 56 (2009): 247–66.

Fuchtbauer, H., and T. Peryt. *The Zechstein Basin with Emphasis on Carbonate Sequences.* Contributions to Sedimentary Geology 9. Stuttgart: Schweizerbart, 1980.

Galloway, W. "Depositional Evolution of the Gulf of Mexico Sedimentary Basin." In *The Sedimentary Basins of the United States and Canada,* 505–49. Amsterdam: Elsevier, 2008.

Garcia-Castellanos, D., F. Estrada, I. Jiménez-Munt, C. Gorini, M. Fernàndez, J. Vergés, and R. De Vicente. "Catastrophic Flood of the Mediterranean after the Messinian Salinity Crisis." *Nature* 462 (2009): 778–81.

Hay, W. W., R. M. DeConto, C. N. Wold, K. M. Wilson, S. Voigt, M. Schulz, A. Wold-Rossby, et al. "Alternative Global Cretaceous Paleogeography." In *Evolution of the Cretaceous Ocean-Climate System: Boulder, Colorado.* Edited by E. Barrera and C. C. Johnson, 1–47. Geological Society of America Special Paper 332. Boulder, CO: Geological Society of America, 1999.

Hay, W. W., A. Migdisov, A. N. Balukhovsky, C. N. Wold, S. Flögel, and E. Söding. "Evaporites and the Salinity of the Ocean during the Phanerozoic: Implications for Climate, Ocean Circulation and Life." *Palaeogeography, Palaeoclimatology, Palaeoecology* 240 (2006): 3–46.

Ho, S. L., and T. Laepple. "Flat Meridional Temperature Gradient in the Early Eocene in the Subsurface Rather Than Surface Ocean." *Nature Geoscience* 9 (2016): 606–10.

Huber, B. T., R. D. Norris, and K. G. MacLeod. "Deep-Sea Paleotemperature Record of Extreme Warmth during the Cretaceous." *Geology* 30 (2002): 123–26.

Jaraula, C.M.B., K. Grice, R. J. Twitchett, M. E. Böttcher, P. LeMetayer, G. Apratim, A. G. Dastidar, and F. L. Opazo. "Elevated pCO2 Leading to End Triassic Extinction, Photic Zone Euxinia and Rising Sea Levels." *Geology* 41 (2013): 955–58.

Jenkyns, H. C. "Geochemistry of Oceanic Anoxic Events." *Geochemistry, Geophysics, Geosystems* 11 (2010): Q03004. doi:10.1029/2009GC002788.

Kasprak, A., J. Sepulveda, R. Price-Waldman, K. Williford, S. Schoepfer, J. Haggart, P. Ward, R. Summons, and J. Whiteside. "Episodic Photic Zone Euxinia in the Northeastern Panthalassic Ocean during the End-Triassic Extinction." *Geology* 43 (2015): 307–10.

Keller, C., P. Hochuli, H. Weissert, S. Bernasconi, M. Giorgioni, and T. Garcia. "A Volcanically Induced Climate Warming and Floral Change Preceded the Onset of OAE1a (Early Cretaceous)." *Palaeogeography, Palaeoclimatology, Palaeoecology* 305 (2011): 43–49.

Larrasoaña, J. C., A. P. Roberts, and E. J. Rohling. "Dynamics of Green Sahara Periods and Their Role in Hominin Evolution." *PLoS ONE*, 8 (2013): e76514. doi:10.1371/journal.pone.0076514.

Leckie, R., T. Bralower, and R. Cashman. "Oceanic Anoxic Events and Plankton Evolution: Biotic Response to Tectonic Forcing during the Mid-Cretaceous." *Paleoceanography* 17 (2002): 13.1–13.29. doi:10.1029/2001PA000623.

"The Mesozoic Era 1. The Mesozoic Era of the Phanerozoic Eon: 251 to 65.5 Million Years Ago." *Palaeos*. Last modified January 19, 2008. http://palaeos.com/mesozoic/mesozoic.htm.

"Messinian Salinity Crisis." *Wikipedia*. Accessed February 8, 2017. https://en.wikipedia.org/wiki/Messinian_salinity_crisis.

Motani, R., D. -Y. Jiang, G. -B. Chen, A. Tintori, O. Rieppel, C. Ji, J. -D. Huang. "A Basal Ichthyosauriform with a Short Snout from the Lower Triassic of China." *Nature* 517 (2015): 485–88.

Monteiro, F., R. Pancost, A. Ridgwell, and Y. Donnadieu. "Nutrients as the Dominant Control on the Spread of Anoxia and Euxinia across the Cenomanian-Turonian Oceanic Anoxic Event (OAE2): Model-Data Comparison." *Paleoceanography* 27 (2012): PA4209. doi:10.1029/2012PA002351.

"New Consensus on Messinian Salinity Crisis: CIESM Workshop 33." *CIESM*. Last modified February 21, 2008. http://www.ciesm .org/news/ciesm/p200208.htm.

NOAA Great Lakes Environmental Research Laboratory (GLERL). "Harmful Algal Blooms in the Great Lakes: What They Are and How They Can Affect Your Health." Ann Arbor, MI: NOAA GLERL, n.d., accessed February 8, 2017. http://www.glerl.noaa .gov/pubs/brochures/bluegreenalgae_factsheet.pdf.

O'Leary, M. A., J. I. Bloch, J. J. Flynn, T. J. Gaudin, A. Giallombardo, N. P. Giannini, S. L. Goldberg, et al. "The Placental Mammal Ancestor and the Post-K-Pg Radiation of Placentals." *Science* 339 (2013): 662–67.

Poulsen, C., E. Barron, M. Arthur, and W. Peterson. "Response of the Mid-Cretaceous Global Oceanic Circulation to Tectonic and CO_2 Forcings." *Paleoceanography* 16 (2001): 576–92.

Price, G. D., R. J. Twitchett, J. R. Wheeley, and G. Buono. "Isotopic Evidence for Long-Term Warmth in the Mesozoic." *Scientific Reports* 3 (2013): 1438. doi:10.1038/srep01438.

Qin, W., L. T. Carlson, E. V. Armbrust, A. H. Devol, J. W. Moffett, D. A. Stahl, and A. E. Ingalls. "Confounding Effects of Oxygen and Temperature on the TEX_{86} Signature of Marine Thaumarchaeota." *Proceedings of the National Academy of Sciences of the USA* 112 (2015): 10979–84.

Rohling, E. J. "Review and New Aspects Concerning the Formation of Mediterranean Sapropels." *Marine Geology* 122 (1994): 1–28.

Rohling, E. J. "The Dark Secret of the Mediterranean: A Case History in Past Environmental Reconstruction." *Highstand.org*. Last modified January 7, 2002. http://www.highstand.org/erohling /DarkMed/dark-title.html.

Rohling, E. J., G. L. Foster, K. M. Grant, G. Marino, A. P. Roberts, M. E. Tamisiea, and F. Williams. "Sea-Level and Deep-Sea-Temperature

Variability over the Past 5.3 Million Years." *Nature* 508 (2014): 477–82.

Rohling, E. J., G. Marino, and K. M. Grant. "Mediterranean Climate and Oceanography, and the Periodic Development of Anoxic Events (Sapropels)." *Earth-Science Reviews* 143 (2015): 62–97.

Rohling, E. J., R. Schiebel, and M. Siddall. "Controls on Messinian Lower Evaporite Cycles in the Mediterranean." *Earth and Planetary Science Letters* 275 (2008): 165–71.

Sellwood, B., and P. Valdes. "Mesozoic Climates: General Circulation Models and the Rock Record." *Sedimentary Geology* 190 (2006): 269–87.

Sijp, W., A. von der Heydt, H. Dijkstra, S. Flögel, P. Douglas, and P. Bijl. "The Role of Ocean Gateways on Cooling Climate on Long Time Scales." *Global and Planetary Change* 119 (2014): 1–22.

Takashima, R., H. Nishi, B. T. Huber, and R. M. Leckie. "Greenhouse World and the Mesozoic Ocean." *Oceanography* 19 (2006): 82–92.

Tanner, L. H., S. G. Lucas, and M. G. Chapman. "Assessing the Record and Causes of Late Triassic Extinctions." *Earth-Science Reviews* 65 (2004): 103–39.

Torsvik, T., S. Rousse, C. Labails, and M. Smethurst. "A New Scheme for the Opening of the South Atlantic Ocean and the Dissection of an Aptian Salt Basin." *Geophysical Journal International* 177 (2009): 1315–33.

Trabucho Alexandre, J., E. Tuenter, G. Henstra, K. van der Zwan, R. van de Wal, H. Dijkstra, and P. de Boer. "The Mid-Cretaceous North Atlantic Nutrient Trap: Black Shales and OAEs." *Paleoceanography* 25 (2010): PA4201. doi:10.1029/2010PA001925.

Van de Schootbrugge, B., and P. B. Wignall. "A Tale of Two Extinctions: Converging End-Permian and End-Triassic Scenarios." *Geological Magazine* 153 (2016): 332–54.

Warren, J. "Evaporites through Time: Tectonic, Climatic and Eustatic Controls in Marine and Nonmarine Deposits." *Earth-Science Reviews* 98 (2010): 217–68.

Westerhold, T., U. Röhl, T. Frederichs, S. M. Bohaty, and J. C. Zachos. "Astronomical Calibration of the Geological Timescale: Closing the Middle Eocene Gap." *Climate of the Past* 11 (2015): 1181–95.

Whiteside, J. H., P. E. Olsen, T. Eglinton, M. E. Brookfield, and R. N. Sambrotto. "Compound-Specific Carbon Isotopes from Earth's Largest Flood Basalt Eruptions Directly Linked to the End-Triassic Mass Extinction." *Proceedings of the National Academy of Sciences of the USA* 107 (2010): 6721–25.

Wilson, H. "Extensional Evolution of the Gulf of Mexico Basin and the Deposition of Tertiary Evaporites." *Journal of Petroleum Geology* 26 (2003): 403–28.

Woo, K., T. Anderson, L. Railsback, and P. Sandberg. "Oxygen Isotope Evidence for High-Salinity Surface Seawater in the Mid-Cretaceous Gulf of Mexico: Implications for Warm, Saline Deepwater Formation." *Paleoceanography* 7 (1992): 673–85.

Wortmann, U. G., and B. M. Chernyavsky. "Effect of Evaporite Deposition on Early Cretaceous Carbon and Sulphur Cycling." *Nature* 446 (2007): 654–56.

KEY SOURCES FOR CHAPTER 7

Appenzeller, T. "Great Green North: Was the Icy Arctic Once a Warm Soup of Life?" *National Geographic*. Accessed February 8, 2017. http://ngm.nationalgeographic.com/ngm/0505/resources _geo.html.

Barker, S., G. Knorr, R. L. Edwards, F. Parrenin, A. E. Putnam, L. C. Skinner, E. Wolff, and M. Ziegler. "800,000 Years of Abrupt Climate Variability." *Science* 334 (2011): 347–51.

Beerling, D., and D. Royer. "Convergent Cenozoic CO_2 History." *Nature Geoscience* 4 (2011): 418–20.

Bijl, P. K., A. J. Houben, S. Schouten, S. M. Bohaty, A. Sluijs, G. J. Reichart, J. S. Sinninghe Damsté, and H. Brinkhuis. "Transient Middle Eocene Atmospheric CO_2 and Temperature Variations." *Science* 330 (2010): 819–21.

Brinkhuis, H., S. Schouten, M. E. Collinson, A. Sluijs, J. S. Sinninghe Damsté, G. R. Dickens, M. Huber, et al. "Episodic Fresh Surface Waters in the Eocene Arctic Ocean." *Nature* 441 (2006): 606–9.

Broecker, W., J. Yu, and A. Putnam. "Two Contributors to the Glacial CO_2 Decline." *Earth and Planetary Science Letters* 429 (2015): 191–96.

Coxall, H. K., P. A. Wilson, H. Pälike, C. H. Lear, and J. Backman. "Rapid Stepwise Onset of Antarctic Glaciation and Deeper

Calcite Compensation in the Pacific Ocean." *Nature* 433 (2005): 53–57.

Crucifix, M. "Oscillators and Relaxation Phenomena in Pleistocene Climate Theory." *Philosophical Transaction of the Royal Society of London, A* 370 (2012): 1140–65.

DeConto, R. M., and D. Pollard. "Rapid Cenozoic Glaciation of Antarctica Induced by Declining Atmospheric CO_2." *Nature* 421 (2003): 245–49.

DeConto, R. M., D. Pollard, P. A. Wilson, H. Pälike, C. H. Lear, and M. Pagani. "Thresholds for Cenozoic Bipolar Glaciation." *Nature* 455 (2008): 652–56.

Diester-Haass, L., and R. Zahn. "Eocene-Oligocene Transition in the Southern Ocean: History of Water Mass Circulation and Biological Productivity." *Geology* 24 (1996): 163–66.

Elderfield, H., P. Ferretti, M. Greaves, S. Crowhurst, I. N. McCave, D. Hodell, and A. M. Piotrowski. "Evolution of Ocean Temperature and Ice Volume through the Mid-Pleistocene Climate Transition." *Science* 337 (2012): 704 –9.

Eldrett, J. S., I. C. Harding, P. A. Wilson, E. Butler, and A. P. Roberts. "Continental Ice in Greenland during the Eocene and Oligocene." *Nature* 446 (2007): 176–79.

Feakins, S., S. Warny, and J. Lee. "Hydrologic Cycling over Antarctica during the Middle Miocene Warming." *Nature Geoscience* 5 (2012): 557–60.

Foster, G. L., C. Lear, and J. Rae. "CO_2 and Climate Closely Linked during the Middle Miocene (11 to 17 Million Years Ago)." *Earth System* (blog). July 31, 2012. http://descentintotheicehouse.org .uk/826/.

Foster, G. L., C. Lear, and J. Rae. "The Evolution of pCO_2, Ice Volume and Climate during the Middle Miocene." *Earth and Planetary Science Letters* 341–44 (2012): 243–54.

Foster, G. L., and E. J. Rohling. "Relationship between Sea Level and Climate Forcing by CO_2 on Geological Timescales." *Proceedings of the National Academy of Sciences of the USA* 110 (2013): 1209–14.

Foster, G., D. Royer, and D. Lunt. "Past and Future CO_2." *Skeptical-Science.com*. Last modified May 1, 2014. http://www.skeptical science.com/print.php?n=2502.

Goldner, A., N. Herold, and M. Huber. "Antarctic Glaciation Caused Ocean Circulation Changes at the Eocene-Oligocene Transition." *Nature* 511 (2014): 574–77.

Grant, K. M., E. J. Rohling, C. Bronk Ramsey, H. Cheng, R. L. Edwards, F. Florindo, D. Heslop, et al. "Sea-Level Variability over Five Glacial Cycles." *Nature Communications* 5 (2014): 5076. doi:10.1038 /ncomms6076, 2014.

Greenop, R., G. Foster, P. Wilson, and C. Lear. "Middle Miocene Climate Instability Associated with High-Amplitude CO_2 Variability." *Paleoceanography* 29 (2014): 845–53.

Houben, A.J.P., P. K. Bijl, J. Pross, S. M. Bohaty, S. Passchier, C. E. Stickley, U. Röhl, et al. "Reorganization of Southern Ocean Plankton Ecosystem at the Onset of Antarctic Glaciation." *Science* 340 (2013): 341–44.

"Ice Core." *Wikipedia*. Accessed February 8, 2017. https://en.wikipe dia.org/wiki/Ice_core.

Kender, S., J. Yu, and V. L. Peck. "Deep Ocean Carbonate Ion Increase during Mid Miocene CO_2 Decline." *Scientific Reports* 4 (2014): 4187. doi:10.1038/srep04187.

Lisiecki, L., and M. Raymo. "A Pliocene-Pleistocene Stack of 57 Globally Distributed Benthic $\delta^{18}O$ Records." *Paleoceanography* 20 (2005): PA1003. doi:10.1029/2004PA001071.

Martínez-Botí, M. A., G. L. Foster, T. B. Chalk, E. J. Rohling, P. F. Sexton, D. J. Lunt, R. D. Pancost, M.P.S. Badger, and D. N. Schmidt. "Plio-Pleistocene Climate Sensitivity Evaluated Using High-Resolution CO_2 Records." *Nature* 518 (2015): 49–53.

Montes, C., A. Cardona, C. Jaramillo, A. Pardo, J. C. Silva, V. Valencia, C. Ayala, et al. "Middle Miocene Closure of the Central American Seaway." *Science* 348 (2015): 226–29.

"An Online Guide to Sequence Stratigraphy." *University of Georgia Stratigraphy Lab*. Accessed February 8, 2017. http://strata.uga.edu /sequence/index.html.

Pagani, M., M. Huber, Z. Liu, S. M. Bohaty, J. Henderiks, W. Sijp, S. Krishnan, and R. M. DeConto. "The Role of Carbon Dioxide during the Onset of Antarctic Glaciation." *Science* 334 (2011): 1261–64.

Pearson, P. N., G. L. Foster, and B. S. Wade. "Atmospheric Carbon Dioxide through the Eocene-Oligocene Climate Transition." *Nature* 461 (2009): 1110–13.

Pekar, S., R. DeConto, and D. Harwood. "Resolving a Late Oligocene Conundrum: Deep-Sea Warming and Antarctic Glaciation." *Palaeogeography, Palaeoclimatology, Palaeoecology* 231 (2006): 29–40.

Rohling, E. J., and S. Cooke. "Stable Oxygen and Carbon Isotope Ratios in Foraminiferal Carbonate." In *Modern Foraminifera*. Edited by B. K. Sen Gupta, 239–58. Dordrecht, the Netherlands: Kluwer Academic, 1999.

Rohling, E. J., G. L. Foster, K. M. Grant, G. Marino, A. P. Roberts, M. E. Tamisiea, and F. Williams. "Sea-Level and Deep-Sea-Temperature Variability over the Past 5.3 Million Years." *Nature* 508 (2014): 477–82.

Rohling, E. J., M. Medina-Elizalde, J. G. Shepherd, M. Siddall, and J. D. Stanford. "Sea Surface and High-Latitude Temperature Sensitivity to Radiative Forcing of Climate over Several Glacial Cycles." *Journal of Climate* 25 (2012): 1635–56.

Ruddiman, W. F. "A Paleoclimatic Enigma?" *Science* 328 (2010): 838–39.

Siddall, M., E. J. Rohling, T. Blunier, and R. Spahni. "Patterns of Millennial Variability over the Last 500 ka." *Climate of the Past* 6 (2010): 295–303.

Sigman, D. M., and E. A. Boyle. "Glacial/Interglacial Variations in Atmospheric Carbon Dioxide." *Nature* 407 (2000): 859–69.

Sigman, D. M., M. P. Hain, and G. H. Haug. "The Polar Ocean and Glacial Cycles in Atmospheric CO_2 Concentration." *Nature* 466 (2010): 47–55.

Snedden, J. W., and C. Liu. "A Compilation of Phanerozoic Sea-Level Change, Coastal Onlaps and Recommended Sequence Designations." *Search and Discovery* August 20, 2010: 40594. http://www.searchanddiscovery.com/pdfz/documents/2010/40594snedden/ndx_snedden.pdf.html.

Stocker, T., and S. Johnsen. "A Minimum Thermodynamic Model for the Bipolar Seesaw." *Paleoceanography* 18 (2003): 1087. doi:10.1029/2003PA000920.

Teschner, C. "Reconstructing the Plio-Pleistocene Evolution of the Water Mass Exchange and Climate Variability in the Nordic Seas and North Atlantic Ocean." *Christian-Albrechts-Universität zu Kiel*. Dissertation [abstract]. January 10, 2014. http://macau.uni-kiel.de/receive/dissertation_diss_00013996.

Whaley, J. "The *Azolla* Story: Climate Change and Arctic Hydrocarbons." *Geo ExPro* 4, no. 4 (2007): 66–72.

Yu, J., R. F. Anderson, and E. J. Rohling. "Deep Ocean Carbonate Chemistry and Glacial-Interglacial Atmospheric CO_2 Changes." *Oceanography* 27 (2014): 16–25.

Zachos, J. C., G. R. Dickens, and R. E. Zeebe. "An Early Cenozoic Perspective on Greenhouse Warming and Carbon-Cycle Dynamics." *Nature* 451 (2008): 279–83.

Zachos, J. C., M. Pagani, L. Sloan, E. Thomas, and K. Billups. "Trends, Rhythms, and Aberrations in Global Climate 65 Ma to Present." *Science* 292 (2001): 686–93.

KEY SOURCES FOR CHAPTER 8 AND EPILOGUE

Abbot, D., and E. Tziperman. "A High-Latitude Convective Cloud Feedback and Equable Climates." *Quarterly Journal of the Royal Meteorological Society* 134 (2008): 165–85.

Adams, S., F. Baarsch, A. Bondeau, D. Coumou, R. Donner, K. Frieler, B. Hare, et al. *Turn Down the Heat: Climate Extremes, Regional Impacts, and the Case for Resilience—Full Report*. Washington, DC: World Bank, 2013. http://documents.worldbank.org/curated/en/2013/06/17862361/turn-down-heat-climate-extremes-regional-impacts-case-resilience-full-report.

"Albedo." *Encyclopedia of Earth*. Accessed February 8, 2017. http://www.eoearth.org/view/article/149954/.

Burke, M., S. M. Hsiang, and E. Miguel. "Global Non-linear Effect of Temperature on Economic Production." *Nature* 527 (2015): 235–39.

Ceballos, G., P. Ehrlich, A. Barnosky, A. Garcia, R. Pringle, and T. Palmer. "Accelerated Modern Human-Induced Species Losses: Entering the Sixth Mass Extinction." *Science Advances* 1 (2015): e1400253. doi:10.1126/sciadv.1400253.

Channell, J., E. Curmi, O. Nguyen, E. Prior, A. R. Syme, H. R. Jansen, E. Rahbari, E. L. Morse, S. M. Kleinman, and T. Kruger. *Energy Darwinism II: Why a Low Carbon Future Doesn't Have to Cost the Earth*. N.p.: Citi GPS, 2015. https://ir.citi.com/hsq32Jl1m4aIzicMqH8sBkPnbsqfnwy4Jgb1J2kIPYWIw5eM8yD3FY9VbGpK%2Baax.

"The Extinction Crisis." *Center for Biological Diversity*. Accessed February 8, 2017. http://www.biologicaldiversity.org/programs/bio diversity/elements_of_biodiversity/extinction_crisis/.

Forsberg, R., V. Barletta, J. Levinsen, J. Nilsson, and L. Sørensen. "Mass Loss of Greenland and Antarctica from GRACE, IceSat and Cryo-Sat." N.p.: Technical University of Denmark (DTU) Space, National Space Institute, n.d. http://congrexprojects.com/docs/12c20_docs2 /2-grace_esa-clic_forsberg.pdf?sfvrsn=2.

Foster, G. L., and E. J. Rohling. "Relationship between Sea Level and Climate Forcing by CO_2 on Geological Timescales." *Proceedings of the National Academy of Sciences of the USA* 110 (2013): 1209–14.

Hansen, J., P. Kharecha, M. Sato, V. Masson-Delmotte, F. Ackerman, D. Beerling, P. J. Hearty, et al. "Assessing 'Dangerous Climate Change': Required Reduction of Carbon Emissions to Protect Young People, Future Generations and Nature." *PLoS ONE* 8 (2013): e81648. doi:10.1371/journal.pone.0081648.

Hönisch, B., A. Ridgwell, D. N. Schmidt, E. Thomas, S. J. Gibbs, A. Sluijs, R. Zeebe, et al. "The Geological Record of Ocean Acidification." *Science* 335 (2012): 1058–63.

Howat, I., C. Porter, M. Noh, B. Smith, and S. Jeong. "Brief Communication: Sudden Drainage of a Subglacial Lake beneath the Greenland Ice Sheet." *The Cryosphere* 9 (2015): 103–8.

Hsiang, S. M., M. Burke, and E. Miguel. "Quantifying the Influence of Climate on Human Conflict." *Science* 341 (2013): 1235367. doi:10.1126/science.1235367.

Isaac, J., and S. Turton. "Expansion of the Tropics: Evidence and Implications." *State of the Tropics*. Essay 5 [2014]. Accessed February 8, 2017. http://stateofthetropics.org/state-of-the-tropics-the-essays.

Kirk-Davidoff, D., and J. Lamarque. "Maintenance of Polar Stratospheric Clouds in a Moist Stratosphere." *Climate of the Past* 4 (2008): 69–78.

Kossin, J. P., K. A. Emanuel, and G. A. Vecchi. "The Poleward Migration of the Location of Tropical Cyclone Maximum Intensity." *Nature* 509 (2014): 349–52.

Laliberté, F., J. Zika, L. Mudryk, P. J. Kushner, J. Kjellsson, and K. Döös. "Constrained Work Output of the Moist Atmospheric Heat Engine in a Warming Climate." *Science* 347 (2015): 540–43.

Lelieveld, J., J. S. Evans, M. Fnais, D. Giannadaki, and A. Pozzer. "The Contribution of Outdoor Air Pollution Sources to Premature Mortality on a Global Scale." *Nature* 525 (2015): 367–71.

"Low Sea Ice Extent Continues in Both Poles." *National Snow & Ice Data Center.* Accessed February 8, 2017. http://nsidc.org/arctic seaicenews/2017/01/low-sea-ice-extent-continues-in-both-poles/.

McGranahan, G., D. Balk, and B. Anderson. "The Rising Tide: Assessing the Risks of Climate Change and Human Settlements in Low Elevation Coastal Zones." *Environment and Urbanisation* 19 (2007): 17–37.

Nagelkerken, I., and S. D. Connell. "Global Alteration of Ocean Ecosystem Functioning Due to Increasing Human CO_2 Emissions." *Proceedings of the National Academy of Sciences of the USA* 112 (2015): 13272–77.

Nicholls, R. J., N. Marinova, J. A. Lowe, S. Brown, P. Vellinga, D. de Gusmão, J. Hinkel, and R.S.J. Tol. "Sea-Level Rise and Its Possible Impacts Given a 'Beyond 4°C World' in the Twenty-First Century." *Philosophical Transactions of the Royal Society of London, A* 13 (2011): 161–81.

Piana, M. E. "Polar Stratospheric Clouds." *Harvard John A. Paulson School of Engineering and Applied Sciences.* Accessed February 9, 2017. http://www.seas.harvard.edu/climate/eli/research/equable /psc.html.

Pollard, D., R. DeConto, and R. Alley. "Potential Antarctic Ice Sheet Retreat Driven by Hydrofracturing and Ice Cliff Failure." *Earth and Planetary Science Letters* 412 (2015): 112–21.

Rahmstorf, S., J. Box, G. Feulner, M. Mann, A. Robinson, S. Rutherford, and E. Schaffernicht. "Exceptional Twentieth-Century Slowdown in Atlantic Ocean Overturning Circulation." *Nature Climate Change* 5 (2015): 475–80.

Richter-Menge, J., J. E. Overland, and J. T. Mathis (eds.). *Arctic Report Card 2016.* noaa.gov, 2016. ftp://ftp.oar.noaa.gov/arctic/documents /ArcticReportCard_full_report2016.pdf.

Ritchie, E. "Extinction: Just How Bad Is It and Why Should We Care?" *Conversation*, May 1, 2013. http://theconversation.com/extinction -just-how-bad-is-it-and-why-should-we-care-13751.

Rohling, E. "Without a Longer-Term View, the Paris Agreement Will Lock In Warming for Centuries." *Conversation*, August 23, 2016. https://theconversation.com/without-a-longer-term-view-the-par is-agreement-will-lock-in-warming-for-centuries-64169.

Rose, B., and D. Ferreira. "Ocean Heat Transport and Water Vapor Greenhouse in a Warm Equable Climate: A New Look at the Low Gradient Paradox." *Journal of Climate* 26 (2013): 2117–36.

Seidel, D. J., Q. Fu, W. J. Randel, and T. J. Reichler. "Widening of the Tropical Belt in a Changing Climate." *Nature Geoscience* 1 (2008): 21–24.

Sijp, W., A. von der Heydt, H. Dijkstra, S. Flögel, P. Douglas, and P. Bijl. "The Role of Ocean Gateways on Cooling Climate on Long Time Scales." *Global and Planetary Change* 119 (2014): 1–22.

Sterner, T. "Economics: Higher Costs of Climate Change." *Nature* 527 (2015): 177–78. doi:10.1038/nature15643.

Velicogna, I., T. Sutterley, and M. van den Broeke. "Regional Acceleration in Ice Mass Loss from Greenland and Antarctica Using GRACE Time-Variable Gravity Data." *Geophysical Research Letters* 41 (2014): 8130–37.

Willis, M. J., B. G. Herried, M. G. Bevis, and R. E. Bell. "Recharge of a Subglacial Lake by Surface Meltwater in Northeast Greenland." *Nature* 518 (2015): 223–27.

Winkelmann, R., A. Levermann, A. Ridgwell, and K. Caldeira. "Combustion of Available Fossil Fuel Resources Sufficient to Eliminate the Antarctic Ice Sheet." *Science Advances* 1 (2015): e1500589. doi:10.1126/sciadv.1500589.

WWF. *Living Blue Planet Report: Species, Habitats and Human Well-Being.* Gland, Switzerland: WWF, 2015. https://www.worldwildlife .org/publications/living-blue-planet-report-2015.

INDEX

Note: Page numbers in *italics* refer to figures in the text.

abyssal plains, 22–23

ACC. *See* Antarctic Circumpolar Current

acidification. *See* ocean acidification

Adhemar, Joseph, 62

aerosols, *58*, 92, 175, 202

Africa, 6, 94, 132, 186, 189; green Sahara, 147, 149–50; monsoons, 147–51; and plate tectonics, 24, 28, *29*, *30*, 127–28, 130, 156, 158

Agulhas Current, 186, 189

albedo effect, 57–60, *58*, 202; and cloud cover, 34, 57, *58*, 92; ice-albedo effect, *58*, 58–60, 65, 80, 88, 90, 167, 175, 180, 202, 209; and icehouse conditions, 167; land-surface-albedo effect, 80, 202, 204; and location of continents and oceans, 79–80; and snowball Earth periods, 87–88; vegetation-albedo effect, 60, 175, 180, 202, 204

algae, 46, 48, 51, 91–92, 94, 98, 102, 103, 122, 135, 140. *See also* planktonic calcifiers

Alps, 73, 74, 164

ammonoids, *3*, 106, 109, 110, 119

amphibians, *3*, 49

Andes, 73, 74, 80, 163, 203

animals, *3*; early animals, 48–49, 97–99; and icehouse conditions, 166; Mesozoic animals, 126–30; and oxygen, 97–99. *See also* extinction events; life; *specific types*

anoxic and low-oxygen conditions, 5, 44–45, 125, 133–53; and *Azolla* event, 164; and basin restriction, 140, 146, 161; "dead zones," 5, 44, 140; and end-Permian extinction event, 119, 194; and injection of external

carbon into atmosphere-ocean-biosphere system, 110, 114, 125, 133; in the Mediterranean, 144–53; ocean anoxic events, 77–78, 133–53; processes that deplete oxygen, 138–42, 146, 151–52, 164; time scale for, 142, 146

Antarctica: CO_2 threshold for glaciation, 166, 174; continental ice sheets, 59, 165–68, 191, 209; and greenhouse climates, 122, 162; and ice caps, 203; and ice core studies, 178–81, 185, 187; and icehouse conditions, 40–41, 165, 174; and ice volume and sea-level changes, 175; and plate tectonics, 28, *29*, *30*, 127, 164, 165; and seesaw of heat redistribution between hemispheres, 185–88, *188*

Antarctic Circumpolar Current (ACC), 36, 166, 187

Aral Sea, 159

Arctic Ocean, 41, 122, 209; *Azolla* event, 164

Arrhenius, Gustaf, 8

arthropods, *3*, 76, 96

Asia, 132, 213; continental shelves, 20–21; and glaciation, 174, 175; monsoons, 37; mountains, 24, 73, 74, 80, 164, 168; and plate tectonics, 24, *29*, *30*, 91, 127–28, 130, 165

asteroid and comet impacts, 18, 31–32, 80–81, 94, 128–29

astronomical climate cycles, 57, 62–67, 147–49, 166, 175; and ice ages, 175, 176, 184, 191. *See also* Sun

Atlantic Ocean, 24, 104, 123; and seesaw of heat redistribution between hemispheres,